# And Soon I
# Heard a Roaring
# Wind

**Also by Bill Streever**

HEAT:
*Adventures in the World's Fiery Places*

COLD:
*Adventures in the World's Frozen Places*

# And Soon I Heard a Roaring Wind

## A Natural History of Moving Air

# BILL STREEVER

Little, Brown and Company

*New York  Boston  London*

Little, Brown and Company
Hachette Book Group
1290 Avenue of the Americas, New York, NY 10104
littlebrown.com

First Edition: July 2016

Little, Brown and Company is a division of Hachette Book Group, Inc.
The Little, Brown name and logo are trademarks of Hachette Book Group, Inc.

The publisher is not responsible for websites (or their content) that are not
owned by the publisher.

The Hachette Speakers Bureau provides a wide range of authors for speaking
events. To find out more, go to hachettespeakersbureau.com or call (866)
376-6591.

ISBN 978-0-316-41060-1

LCCN 2016930854

10 9 8 7 6 5 4 3 2 1

RRD-C

Printed in the United States of America

Can I dedicate a book to a dead stranger? If so, I humbly dedicate *And Soon I Heard a Roaring Wind* to the scientist and pacifist Lewis Fry Richardson, an admirably intelligent and principled man. If not, I dedicate this book to my wife and co-captain, Lisanne Aerts, who retains a healthy fear of strong winds, and my son, Ish Streever, who shares my love of words and my fascination with the natural world.

# CONTENTS

Introduction: Before Departure                          3

*Chapter 1:* The Voyage                                15

*Chapter 2:* The Forecast                              36

*Chapter 3:* Theorists                                 65

*Chapter 4:* Initial Conditions                        84

*Chapter 5:* The Numbers                              110

*Chapter 6:* The Model                                133

*Chapter 7:* The Computation                          162

*Chapter 8:* Chaos                                    193

*Chapter 9:* The Ensemble                             224

*Chapter 10:* Afloat in the Candle's Light            249


Acknowledgments                                       267

Notes                                                 271

Index                                                 301

And soon I heard a roaring wind:
It did not come anear;
But with its sound it shook the sails,
That were so thin and sere.

—Samuel Taylor Coleridge,
*The Rime of the Ancient Mariner,* 1798


It is absurd to suppose that the air which surrounds us becomes wind simply by being in motion.

—Aristotle, *Meteorologica,* about 350 BC

# And Soon I Heard a Roaring Wind

*Wood-engraved illustration by Gustave Doré from an edition of* The Rime of the Ancient Mariner *published in 1877 in Germany.*

# INTRODUCTION

## *Before Departure*

Aboard the sailing yacht *Rocinante,* the north wind shrieks through rigging. It holds flags taut. It pushes against the boat's hull, and the boat in turn pulls on her dock lines, straining. Dark clouds gallop across the sky.

Eager to begin our voyage but waiting for the norther to blow itself out, I find my mind consumed by wind. It is with me during the day, before I sleep at night, when I awake in the morning. At times, it occupies my dreams.

I do not contemplate mild breezes.

I think of the storm of 1900, with thousands dead in Texas, their bodies buried in rubble and strewn along railroad tracks and floating at sea. I think of the Great Hurricane of 1780 sweeping away more than twenty thousand souls in the Caribbean. I think of a steamship in 1857 encountering an unnamed storm off South Carolina, her crew and passengers in a bucket brigade frantically bailing while the air roared around them, before the sinking vessel took 425 people to their graves. I think of Lawrence Kern in 1930, lifted from the ground by a tornado and found, mortally injured but still alive, a mile away.

I think, too, of Daniel Defoe's storm. Fifteen years before he wrote *Robinson Crusoe,* in a book often described as an early example of journalism, Defoe wrote *The Storm: or, A Collection of the Most Remarkable Casualties and Disasters Which Happen'd in the Late Dreadful Tempest, Both by Sea and Land.*

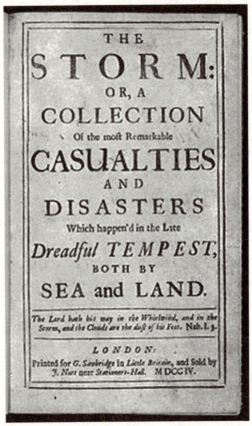

*Daniel Defoe's book, often described as an early example of journalism, did not sell well. (Image from Wikimedia Commons)*

He watched the wind blow day after day for more than a week. And then he watched the wind peak. The storm—though Defoe could not have known this until later, when he collected accounts from witnesses—cut a path three hundred miles wide across England and Wales.

Defoe called it "the greatest, the longest in duration, the widest

4

in extent, of all the Tempests and storms that history gives any account of since." It would become known as the Great Storm of 1703. After three centuries, it is still considered to be England's worst storm.

At first, the wind was not so strong that it could carry a man aloft. If there were no flying debris, a wind like this could be fun. One could lean into it, striking unusual poses. But there was flying debris. Wind-launched missiles killed men and women and children. Defoe himself watched blowing roof tiles hit the ground, embedding themselves, he wrote, "five to eight inches into solid Earth."

As the wind grew in strength, homes shook and threatened to collapse and did collapse, but the prudent and the wise feared the outdoors more than the indoors. "Most People expected the Fall of their Houses," Defoe wrote. "And yet in this general Apprehension, no body durst quit their tottering Habitations; for whatever the Danger was within doors, 'twas worse without."

Even indoors, falling debris claimed lives. One account reported the bishop of Bath and Wells leaping from his bed as the room around him shook. He "made toward the door, where he was found with his Brains dash'd out."

More than one person, feeling a house tremble in the wind, reported an earthquake. But it was the wind.

Impartial, moving air shook churches just as it shook homes. Church bells, unattended, rang. Seven steeples blew down. Where steeples survived, many lost their tops or parts of their tops, sending tiles and bricks and wrought iron crashing down.

Chimneys collapsed. The wind tore rolls of lead sheathing from roofs and sent them on their way. Trees with trunks three feet and more in diameter succumbed. The storm triumphed over oaks and elms and apple trees, not in the hundreds but in the tens of thousands. They were lifted out of the ground, roots intact, and sent flying over fences and hedgerows. They were snapped off at chest height. Their trunks were twisted in ways unknown to carpenters.

Defoe reported the loss of four hundred windmills. Some tumbled, their heavy anchor posts suddenly surrendering, failing under the load. Others burned, their wooden innards ignited by the friction of moving parts rubbing one against the other as their blades spun out of control.

Seawater flew inland, carried in raging gusts. One man reported "Froth and Sea moisture six or seven miles up the Country, for at that distance from the Sea, the Leaves of the Trees and Bushes, were as Salt as if they had been dipped in the Sea, which can be imputed to nothing else, but the Violent Wind." The salt water reached pastures twenty-five miles from the nearest windward coast. "The Grass was so salt, the Cattel would not eat for several Days."

On land, the death toll was surprisingly light. "We have reckoned," Defoe reported, "123 People kill'd."

At sea, it was a different story. At a time when the population of England and Wales hovered around five million, an estimated eight thousand sailors died in that storm.

Offshore, waves washed over decks and smashed through portholes. Ships raced down wave faces, crashing into troughs with an impact that jolted planks and beams and tore at the fasteners that held them together. While sailors fought for their lives, their ships moved beneath them like roller coasters. The sailors did not give up easily. They manned pumps. They cut away rigging and masts to reduce windage. They held on or were lost.

Nearer shore, sailors watched as anchors broke free. They tried to guide their ships away from deadly reefs and rocks, where wooden hulls would be reduced to splinters by pounding seas and where they themselves might be battered to death before they drowned.

The worst of the wind came well after sunset. "The night was exceeding dark," one survivor reported. On deck, sailors could not hear the commands of officers. From the same survivor: "Words were no sooner uttered than they were carried away by the Wind."

And this: "We plainly saw the Wind skimming up the Water, as if it had been Sand, carrying it up into the Air."

Ships, filling with water, sank.

The Royal Navy kept complete records. Three hundred and eighty-seven sailors died with the loss of the *Restoration*. Two hundred and twenty went down with the *Northumberland*. Two hundred and six drowned or were otherwise killed with the sinking of the *Sterling Castle*. Two hundred and sixty-nine were gone with the *Mary*.

Records from the fishing and cargo fleets were less complete. A man from Brighthelmston in Sussex sent Defoe a letter. His town, he wrote, looked as if it had been bombarded. And then he listed casualties at sea. "Derick Pain, Junior, Master of the *Elizabeth* Ketch of this Town lost, with all his company. George Taylor, Master of the Ketch call'd the *Happy Entrance*, lost, and his company, excepting Walter Street, who swimming three days on a mast between the Downs and North Yarmouth was at last taken up. Richard Webb, Master of the Ketch call'd the *Richard and Rose* of Brighthelmston, lost, and all his Company."

Boats struck the seabed and sank in shallow bays. They settled to the bottom, sometimes with masts and rigging protruding above the water. Upon those protrusions, seamen clung for survival above the churning sea. In deeper water, survivors hugged floating debris, struggled for breath in the blowing spray and foam, and fought exhaustion on a tirelessly moving landscape of waves.

Offshore from Deal, about seventy-five miles east of London, shipwrecked sailors, momentarily spared, found their way to intertidal flats exposed at low tide. On the sand, they weathered the wind as best they could until the tide, rising, carried them away.

In 1703, there had been no forecast. No warnings were posted before the storm hit. And no one could explain the storm.

"Those Ancient Men of Genius," Defoe wrote, "who rifled Nature by the Torch-Light of Reason even to her very Nudities, have been run a-ground in this unknown Channel; the Wind has blown out the Candle of Reason, and left them all in the Dark."

Defoe, like his contemporaries, blamed the storm on God.

Flash forward more than two hundred years. The British Quaker scientist and World War I ambulance driver Lewis Fry Richardson contemplated weather. Like many of his contemporaries, he believed that wind and weather would abide by the rules of physics. Unlike most of his contemporaries, he thought that the rules of physics behind wind and weather could be described in a series of equations. If that was true, tomorrow's weather could be divined by combining simple measurements of today's weather with equations for Newton's laws of motion and the basic principles of thermodynamics. Forecasting could be done without guesswork, without intuition, without the subjective interpretation of weather maps. Human beings—the ones who understood something about hydrodynamics and thermodynamics, the ones who knew how to manipulate numbers—could calculate the comings and goings of storms, calms, and favorable winds. With mathematics, they could see into the future. In 1922, Richardson published *Weather Prediction by Numerical Process*.

The recipe was, in general terms, straightforward. First, lay down a grid dividing the atmosphere into thousands of cells, wrapping the earth in something like a three-dimensional chess-

*Lewis Fry Richardson, scientist and pacifist, and the first man to calculate the weather. (Image from Wikimedia Commons)*

board. Second, populate each cell with data representing current conditions, with data on things such as wind speed, barometric pressure, and temperature. Third, apply the magic of mathematics to understand how each cell in the grid affects its neighbors over a short period of time—say, six hours. Fourth, run the calculations forward to generate a forecast for tomorrow, and the day after, and the day after that.

Richardson believed that tomorrow's weather was as foreseeable as the positions of the planets. The mathematics involved, in his view, were not especially difficult. They were merely repetitive.

In his book, Richardson dreamed of "a theater, except that the circles and galleries go right round through the space usually occupied by the stage." In his imagined theater, a map of the earth covered the walls. The ceiling held the north pole. Antarctica sat on the floor. Workstations—one for each of the cells of the forecasting grid—packed the theater. At each station, a person hunched over a pencil and paper and a mechanical calculator or a slide rule. During a time before computers, people employed to crunch numbers were called "computers." Their job was, after all, to compute. More specifically, their job was to compute conditions within their tiny workstation cells and to post the results on signs that their neighbors could read. They would then wait to see results posted by their colleagues, and each person—each computer—would use the results posted nearby to run another set of calculations. And each person—each computer—would do exactly the same thing again and again and again, generating results that would stay ahead of the actual weather by a few days.

*Lewis Fry Richardson imagined a vast theater representing the surface of the earth and populated by thousands of workers, each processing the mathematics of the weather in an assigned region and all under the direction of a supervisor in a pulpit guiding the rate of their calculations with beams of light. (Drawing courtesy of Professor Lennart Bengtsson)*

Dead center in Richardson's imagined theater, a pillar rose high above the floor. On top of that pillar sat a pulpit. "In this," wrote Richardson, "sits the man in charge of the whole theater; he is surrounded by several assistants and messengers." Richardson imagined that some groups of computers might be more efficient than others, that they might start to race ahead, so the man in the pulpit had to control the rate of the work. If computations in one part of the theater sprung ahead of the rest, he would turn a red light in that direction. If computations in another part of the theater lagged behind, he would use a blue light. He would maintain, as Richardson put it, "a uniform speed of progress in all parts of the globe."

Richardson, considering the number of calculations required, estimated that he would need sixty-four thousand people to make his imagined theater work—sixty-four thousand computers working frantically, day and night, calculating the weather. He knew that such an effort was impractical, that it would never happen. Having spent many years developing his ideas, he was merely allowing himself to imagine what he knew would never be possible. "After so much hard reasoning," he asked, "may one play with a fantasy?"

But aside from the practical issues of crunching numbers, and despite the initial failure of a small-scale test of his mathematical approach, Richardson was convinced that his concept was sound. He was sure that mathematics would allow humans to see into the future—to see tomorrow's weather today.

His belief in the potential of numerical forecasting was vindicated in 1950 not by an actual theater packed with people scribbling calculations under the glow of red and blue lights, but by an early electronic computer. Following the instructions of a handful of scientists armed with punch cards, the computer produced the first successful numerical forecast, seeing into the future a full twenty-four hours.

Flash forward again, to today. Computers have advanced. Numerical models have advanced. Forecasters gaze further into the future.

By radio, television, computer, or smartphone, a weather forecast is close at hand. Often those forecasts show the future unfolding in one-hour increments. Planning a bicycle ride, a walk, a picnic, or a sail? Taking a paragliding lesson, thinking of hauling a sofa across town in the back of a pickup truck, considering Shakespeare in the Park on a cloudy evening? For the best results, check an hourly forecast.

Modern forecasts rely on complex mathematical models and supercomputers. Now, with some reliability, we see weather, including wind, three days into the future. With somewhat less reliability, we see seven or eight days into the future. With luck, and with substantially less reliability, we sometimes see fourteen days into the future.

But even with supercomputers, even with an understanding of atmospheric physics barely imaginable in 1922, it turns out that wind and weather are not as foreseeable as the positions of the planets. It turns out that Richardson's faith in mathematics and a decipherable deterministic world fell short of its initial promise. Forecasters cannot dispense with experience. They cannot always ignore intuition. Forecasters and those who depend on them cannot blindly rely on a future dictated by mathematics. Forecasters offer a taste of the future, a view of the next few days, a glimpse of uncanny accuracy in most parts of the world, but that picture is not mathematically perfect. Sometimes forecasters get it wrong.

And so, aboard the sailing yacht *Rocinante*, before we cast off, before we begin our voyage, while waiting for the weather to settle, I think about what we do and do not understand about wind. I think about Defoe's storm and a hundred other storms. I think about the scientists who peeled back layers of ignorance, one at a time, most making no more than incremental gains but collectively making sense of our atmosphere and its confusing hyperactivity. I think of the one man whose work showed that the future will be forever uncertain. I think, too, of the beneficiaries of that hard-won science—farmers planting and harvesting crops, pilots planning flights, businesses routing trucks and barges and ships, vacationers, commuters, emergency planners, fishermen, sailors. I think about Wilbur Wright, who watched birds flying in the wind for hours at a time as he designed and redesigned his airplane and who once wrote in his notebook, "No bird soars in a calm." I think of wind turbines springing up like weedy flowers, a sudden energy boom rediscovered, the wind providing enough electricity for eighteen million American homes and jobs for seventy thousand people. And I envision a southern right whale, its twenty-foot-wide tail held high above the water, held up as a sail that catches the wind and pushes the leviathan quietly through the waves.

*Chapter 1*

# THE VOYAGE

Galveston, twenty nautical miles from our dock, was once the most important city in Texas. People called it the Wall Street of the Southwest. Then the storm came. It struck on September 8, 1900, a Saturday. Winds blew at something like 126 knots, or 145 miles per hour. Better estimates would be available if the wind had not carried away the US Weather Bureau's anemometer.

The nameless storm of 1900—naming hurricanes would not become routine until 1953—swept in and moved on very quickly. Storm warning flags were flying on Saturday morning, but Galveston residents, accustomed to storms, went about their business. By Saturday afternoon, the wind had picked up. By dinnertime, it had reached hurricane strength. And by midnight, the wind was dying.

On Sunday morning, sunshine bathed Galveston through clear skies, and the wind blew at a mere seventeen knots, or twenty miles per hour. But the city was ruined.

The *New York Times* covered the story. "Reports are conflicting," the paper said, "but it is known that an appalling disaster has befallen the City of Galveston, where, it is reported, two thousand or more lives have been blotted out." The reporter was correct to write "or more." The storm that carried away the anemometer killed between 6,000 and 12,000 people. It destroyed 3,600 homes.

"The citizens," reported the *Times*, "were all huddled together at the highest points in the center of the town, and consternation and fear reigned almost to the point of madness."

Relief trains attempting to reach Galveston encountered impassable tracks. The prairie across which the tracks ran was covered, according to the *Times,* with "lumber, debris, pianos, trunks, and dead bodies." The third train to try to reach Galveston reported two hundred corpses within sight of the tracks.

On September 11, the *Times* quoted a man named G. L. Russ on conditions after the storm: "I will not attempt to describe the horror of it all; that is impossible. When I left Galveston men armed with Winchester rifles were standing over burying squads and at the point of rifles compelling them to load the corpses on drays, to be hauled to barges, on which they are towed into the Gulf by tugs and tossed into the sea."

Looting was a problem. "Ghouls," the *Times* reported, "stripped dead bodies of jewelry and articles of value."

The people of Galveston rebuilt. Within two years, a three-mile-long seawall was under construction. It grew to a height of seventeen feet. Sand was pumped from the Gulf of Mexico onto the remains of the city, raising its elevation and burying corpses that remained unfound.

The next storm, in 1915, claimed only fifty-three lives. But by then major investments had moved inland, across Galveston Bay. By the 1920s, Galveston had become a place to drink, a version of Bourbon Street on the Texas coast. The hurricane of 1900 had killed thousands and changed the lives of all those who survived.

As we head to sea on this bright morning more than a hundred years after the hurricane, I think about Louisa May Alcott, famous as the author of *Little Women*. "I am not afraid of storms," she wrote, "for I am learning how to sail my ship." I, too, am learning how to sail my ship, but, unlike Alcott, I am scared to death of storms.

Our voyage to understand the wind has begun. We left the dock early in the morning, casting off only to find ourselves stuck in the mud. The north wind had stolen water from the bay, pushing it out to sea, but the bottom near the dock was soft, and we plowed through, cutting our own channel, eager to put Galveston behind us by early afternoon. Trade winds lie ahead, along with calms and squalls and fronts, jet streams, weather forecasters, sand dunes and wind-sculpted rocks, windmills and wind turbines, climate change, and vortexes of spinning air.

And now, not yet two hours off the dock, *Rocinante* cuts through the water at the edge of the Houston Ship Channel. It is just the two of us aboard, my wife and myself, my co-captain and I, a crew of two, both new to sailing but both luxuriating in the feeling of our moving boat. We make five knots, about six miles per hour.

Our boat, *Rocinante,* is a ketch, meaning that she carries two masts, with the aft mast forward of the rudder post. Built in 1965, she has all the problems of age, but also the graceful lines of an older boat, enough so that she attracts compliments. Along with compliments, we field questions about her name. She is named for Don Quixote's horse, an old nag whose delusional rider thinks of her as a steed worthy of a knight.

A muddy tidal current shaves a knot off our speed. A Panamanian flagged tanker, rust-streaked, forces us toward shallow water. We change course, and the foresail needs trimming. My co-captain and I, working together, still learning the ropes, wrap the jib sheet in the wrong direction around a winch. It has to be unwrapped and rewrapped. We get in each other's way. For a moment, our incompetence makes me want to turn back, to come about, to regain the safety and comfort of the dock. But that would be out of character for the crew of *Rocinante*. We have cast off, and we are going.

With a twelve-knot wind, we sail with Houston astern. Before us lies Galveston Island, and then the Gulf of Mexico, and ultimately Guatemala.

We intend to abandon, for a while, certain conveniences, to discard certain comforts, to live for a time closer to the elements, more exposed. In the coming days, we hope to exchange the mud-infused liquid of Galveston Bay for the blue water of the deep Gulf. The plan is to arc east and then south, bearing more or less straight toward Tampa for a while, then toward Key West, and then straight for the tip of the Yucatán. Our first planned landfall is seven or eight days out, or, moving slowly, maybe ten days out, at Mexico's Isla Mujeres, the Island of Women. From there, with the Gulf of Mexico behind us, we will sail south along the coast to Belize and then Guatemala, where we will head upstream into the protected waters of the Río Dulce, the Sweet River, to hide from next summer's hurricanes.

For experienced cruising sailors, this would be a straightforward voyage. Even experienced coastal sailors who had never before ventured beyond sight of land would not be unduly challenged by such a journey. We are neither experienced cruisers nor experienced coastal sailors. We are rank amateurs foolish enough to have purchased a sailboat, ill prepared for such things as broken equipment and storms. Our plans, some would say our lives, depend entirely on the future weather and, in particular, the wind. We are afloat on an erratic ocean, ignorant of the workings of weather forecasts but entirely reliant on their content.

My co-captain is not troubled by this ignorance. She accepts it, reasoning that we are far from the first to make this short journey. In this regard, she lacks my lack of self-confidence. As for me, I have trouble trusting what I do not understand, and so I suffer in that twilight zone between profound anxiety and blossoming fear. But I will not suffer indefinitely. What better place than a sailboat to learn about wind?

I know the story of wind to be complex. It is a story of more than just forecasting. Wind has shaped commerce. It has won and lost wars. It has carved landscapes, made and consumed fortunes, and allowed the first airplane flights.

There is no single thread. No one scientist can be credited with successfully cracking the mysteries of wind. There was no eureka moment in thinking about wind. The story of wind is as much a story of human beings as a story of science.

The term "stormy" can be and has been applied to the history of the science of wind. It is a relatively short history, spanning not much more than a century, but it is a history of misunderstanding, of wrong turns, of heated debates. At times, controversies became very personal. They changed lives. Controversy led, in at least one case, to the suicide of a prominent and otherwise successful man.

In 1704, Daniel Defoe wrote that wind had blown out the candle of reason. Here, afloat at the bottom of this sea of air, worried about weather in the days ahead, I resolve to ignite that candle within myself and with it cast light upon the dark winds that plague my mind. I intend to learn what I can about the wind, about moving air, about its usefulness, about its dangers, about what it carries, about how forecasters forecast, and about why their vision of next week remains, in these days of supercomputers and big science, so cussedly unreliable.

I am not the first to compare the atmosphere to the ocean, to talk of a sea of air. In a 1644 letter describing a mercury barometer, Evangelista Torricelli wrote, *"Noi viviamo sommersi nel fondo d'un pelago, d'aria elementare,"* or something like, "We live submerged at the bottom of an ocean of the element air." The ocean is a liquid and the atmosphere a gas, but both are fluids, substances without fixed shapes, their molecules casually slipping past one another, unencumbered by strong attachments. Both have considerable depth. Both are, under the right conditions, blue. Both have tides and waves. Both flow with complex currents.

The current caressing *Rocinante*'s hull comes from tides, but currents can also be driven by wind. And this role of wind driving water can be reversed. Water, in changing from vapor to liquid and liquid to vapor, can drive wind, and that is part of the story of moving air. Water is not air, and current is not wind, but water and air interact, their fates often married, one affecting the other.

Water is water, and air is air. Among their differences, one stands out. Liquid water, unlike air, is not easily compressed. Molecules of air can be squeezed together. When concentrated, when bunched together, they try to push away from one another, generating pressure. When less concentrated, when the molecules have elbow room, they no longer push away quite so hard. They exert less pressure.

The atmosphere, the sea of air, has pockets of high pressure and pockets of low pressure that are absent from the more or less incompressible water of the world's ocean. The atmosphere has the highs and lows of weather forecasts. And it is these highs and lows that explain wind.

What is wind? It is air moving from pockets of high pressure toward pockets of low pressure, seeking equilibrium, stability, balance. There is nothing difficult about that.

The difficulty comes in understanding why wind seldom moves in a straight line between pockets of high pressure and low pressure, why it never succeeds in reaching equilibrium, why the highs and lows that drive it form and disappear, and why it can provide gentle propulsion one day, threaten peace of mind the next day, and destroy life and property the day after. The difficulty comes in understanding why Lewis Fry Richardson's numerical forecasting does not work as well as one might hope. The difficulty comes in understanding the confusion that arises from the earth's incessant spinning below its atmosphere and from the friction that occurs where moving air meets unyielding ground and trees and buildings and mountains. The difficulty—and the fun—comes in understanding why something that appears at first glance to be so simple can be so wondrously complex.

We did not cast off empty-handed. Aboard, we have canned goods and dry goods, enough to last, without resupply, for thirty days. In our tiny refrigerator, we carry eggs and meat and fresh orange juice, luxury items that have to be consumed soon, before their need for cold storage drains our batteries. We carry two hundred gallons of water. Our medicine chest houses antibiotics and painkillers, bandages and splints, salves and scissors and scalpels. Our toolbox could stock a hardware store. Our spares include gaskets and pulleys and wires and fuses, four kinds of filters for oil and fuel, extra pumps, various lightbulbs, rivets, stainless steel nuts and bolts and screws, and a number of items left behind by *Rocinante's* previous owner that neither of us recognizes but that neither of us is prepared to throw away.

We have radios and a satellite telephone. A radar dome is perched upon our aft mast. Forward of the helm, electronics offer our exact speed and location, along with the speed and location of every ship within ten miles.

Our personal items include clothes for cold weather and for warm, for rain and for shine, as well as inflatable life vests to be worn at all times on deck, each equipped with a satellite beacon.

We have flares, a signaling kite, horns, whistles, mirrors, and an emergency transmitter. We have a life raft, neatly stowed but accessible beneath a bench in the cockpit. Along with it, also neatly stowed but accessible, we have bolt cutters, needed in a demasting to cut away cables that would otherwise tie a fallen mast to the boat. A broken mast, in the kind of winds and seas likely to accompany a demasting, has to be cut free before it batters through the hull, inviting seawater to enter where seawater is the most unwelcome of guests.

Forward, we keep a waterproof bag with energy bars and water bottles, flashlights and an extra radio, passports and money—commodities one might welcome aboard a life raft.

And we have books. *Rocinante* carries a floating library, an assortment of novels and memoirs, travelogues and illustrated keys to marine life, collections of charts and engine manuals, volumes old and new, paper and electronic.

Our knowledge of bolt cutters and life rafts and emergency beacons comes from these books, or mostly from these books. But we also attended a three-day sailing course. We undertook short coastal cruises by way of training. We talked to those more experienced than ourselves. And we talked to those who only talked of casting off, those who had been talking of casting off for so long that their dock lines had failed of old age and been replaced and then replaced again. Conversations among aging dock lines urged us to sea.

To prepare a boat and its novice crew for all eventualities would be to never leave, and so we left not ready for everything, but at least ready for some things. Beyond that, we would trust in luck. With luck, all will go well. By luck I mean with neither a man nor a woman overboard. By luck I mean without breakdowns, without serious groundings, without hitting an oil platform or a shrimper, without being run down by a container ship so massive that the crunching of fiberglass and teak against its hull would fail to register on the ship's bridge. By luck I mean arriving with both masts and all sails intact. But mainly by luck I mean with favorable winds. I mean winds from the right direction and of the right strength. I mean winds that do not blow from dead ahead and that are neither too strong nor too weak. I mean Goldilocks winds.

Globally, well over sixty winds occur frequently enough to be named. There is the sirocco, blowing from the Sahara toward North Africa and southern Europe. There is the Mediterranean's ostro and its bora and its mistral, along with its vendaval, its levanter, and its khamsin. There is India's elephanta and Cuba's bayamo and Chile's puelche. There is the Santa Ana, deadly dry and hot, ripping down the seaward slopes of California's mountains, growing hotter with lost altitude, sometimes fanning wildfires and even driving flames downhill.

From novelist Raymond Chandler in 1938: "It was one of those hot dry Santa Anas that come down through the mountain passes and curl your hair and make your nerves jump and your skin itch. On nights like that every booze party ends in a fight. Meek little wives feel the edge of the carving knife and study their husbands' necks. Anything can happen."

Famously, there are the trade winds, blowing more or less reliably from east to west at the latitude of the Canary Islands and Puerto Rico. The trades propelled the *Santa María*, the *Niña*, and the *Pinta* to the Americas, transporting Christopher Columbus and his men toward San Salvador. For hundreds of years the trade winds drove human migration without conscience, carrying free and enslaved people alike, for better and worse. The trades brought the empty galleons from Spain to the New World, there to be filled with burgled silver and gold and other portable wealth worth hundreds of billions of dollars before sailing north to find the westerlies, the anti-trades, the winds that would carry the ships back to the Old World at latitudes above thirty degrees north.

There is, too, the derecho, a straight-line wind that can exceed eighty knots—eighty nautical miles per hour, or ninety-two statute miles per hour—advancing like a blitzkrieg across the landscape, bringing sudden destruction. There is the tornado, a spinning funnel of wind capable of turning an automobile into a flying projectile. While derechos leave the broken trees of destroyed forests lying with their dead tips pointing in the same general direction, tornadoes leave the broken trees of destroyed forests lying with their dead tips pointing in different directions, confused.

And there is the hurricane, a regional name for a strong tropical cyclone, a spinning weather system, no different from a typhoon except that the hurricane blows in the North Atlantic, the Caribbean, the Gulf of Mexico, and the eastern Pacific, while the typhoon blows in the western Pacific. Both the hurricane and the typhoon circulate around a center of low pressure, a region where the air molecules have elbow room and the wind does not blow. Outside that center, beyond the storm's eye, the air spirals at speeds exceeding sixty-four knots, or seventy-four miles per hour. These are sustained wind speeds—by a standard definition, wind speeds averaged across ten minutes at a height thirty-three feet above the ground, just above midway up the mast of a typical sailing yacht.

The sustained winds of a hurricane, or of any storm, are not its strongest winds. The strongest winds come in gusts. A hurricane's sustained winds and gusts make it difficult or impossible to stand. They lift houses from their foundations. In winds of 115 knots—the winds of strong hurricanes—the simple act of breathing becomes noticeably difficult.

The word "hurricane" came to us from the Spanish, and it came to the Spanish from the Arawak, specifically from the Taino people. The Taino, seafaring Caribbean natives in the time of Columbus, lived scattered across the hurricane heartland, in Cuba, the Bahamas, the Lesser Antilles, and a hundred other low-lying islands.

Columbus himself, before he concluded his business in the

New World, knew both the word *hurakán* and the experience of hurricanes. The regional name and the living experience were also known to Captain John Smith, of Jamestown and Pocahontas fame. In 1627, he described hurricanes for readers in Europe, readers entirely unfamiliar with the winds that often plague the Caribbean and the new colonial coast. The wind, he wrote, comes "with such extremeity that the Sea flies like raine, and the waves so high, they over flow the low grounds by the Sea, in so much, that ships have been driven over tops of high trees there growing, many leagues into the land, and there left."

For now, *Rocinante* rides the remains of the norther south.

By convention, winds are labeled from their direction of origin. A norther—a north wind—blows south. A sea breeze blows from the sea. A land wind blows from the land.

We can count on this north wind for another day or, if our luck holds, two. Soon we will clear the jetties at Galveston Island, passing from the protected bay into the open Gulf. From there we will bear due south through the dense thickets of oil platforms that shadow this coast, putting the worst of them behind us before dusk. But by dusk we need to turn left. We must, as quickly as possible, while the north wind still blows, make easting. The tip of the Yucatán Peninsula, our planned destination, lies south and east of here.

Sailing south from Texas is easy. Sailing east from Texas is not so easy. The north winds come to Texas as fronts and do not last. They clock around to blow from the east. Sailing east from Texas means sailing against the prevailing winds, and sailboats cannot easily sail upwind. When we turn left in the Gulf, the east wind will fall on *Rocinante's* nose, and she will make all the progress of a draft horse climbing a ladder.

Although it is easy enough to make this broad forecast—a

north wind will clock around to blow from the east—the winds
of the northern Gulf of Mexico are not, as a general matter, to
be trusted. They are notorious for changing. They blow from
one direction, stop, blow from another direction, pick up to a
gale, lay down to a flat calm, occasionally spawn the tornado-at-sea
properly known as a waterspout, and for half of each year
threaten the unmitigated terror of a hurricane.

The crew of *Rocinante*—my co-captain and I—long for steady
trade winds, for the winds that blew Columbus to the New
World, for the winds that blow with legendary reliability.

To reach deep into the heart of the trades, we will cross fourteen degrees of latitude. We start now, the day before Thanks-
giving, at the end of the hurricane season. With luck, we will
make easting quickly and then head south, picking up steadier
trades within days. Once we have the trades, we will glide effort-
lessly over clear waters. We will see frolicking dolphins. We will
admire sunsets and sunrises. We will quietly read books while
full sails carry us along.

The twelve-knot Galveston Bay wind brushes my face and
neck. I hear it. I smell what it carries, the airborne stains of
tidal mud and petrochemical plants. But mostly I feel it in *Rocinante's* sails. I feel its power propelling forty thousand pounds of
boat forward, more or less in the direction that I want to go, cutting through the water, leaving behind a rippling wake.

In the 1831 version of what would become his famous wind
scale, Rear Admiral Sir Francis Beaufort of the Royal Navy
called a twelve-knot wind a "fresh breeze." In the modern version of Beaufort's scale, somewhat modified since 1831, his
fresh breeze is considered a moderate breeze. Either way, it tou-
sles what is left of my hair.

It is ordinary wind, not at all miraculous. It is the same wind
that once powered the mills of Persia and Holland and Spain
and that now spins a quarter of a million electricity-generating

turbines. It is the wind exploited by the floating jellyfish-like Portuguese man-of-war, by spiders that fly on silken kites of their own design, by seeds finding their way to isolated oceanic islands. It is the same wind that tumbles tumbleweeds. It once drove farmers, by the thousands, off their land in the American Midwest. It is the carver of stone sculptures called yardangs. It is responsible for the twenty thousand square miles of grainy earth known as the Selima Sand Sheet in southern Egypt and northern Sudan. And it is the very same wind that balances temperatures across the globe, that carries the heat of the equator to the poles and the cold of the poles to the equator, a movement of air that moderates extremes in temperature, extremes that would otherwise leave the planet, for all practical purposes, uninhabitable.

Rear Admiral Sir Francis Beaufort listed thirteen categories of wind in the 1831 version of his famous wind scale. Numbers accompanied the names of the categories, ranging from zero for "calm" to nine for "strong gale" to twelve for "hurricane." He did not give wind speeds themselves, but descriptions of the sea state or the response of sailing vessels likely to be seen for each category. His sailing vessel of reference was the man-of-war, a frigate with a length of around two hundred feet, carrying three masts and perhaps seventy-four cannons, and crowded with upward of five hundred men.

Beaufort's level one, "light air," was "just sufficient to give steerage way," meaning the vessel moved just fast enough to respond to its rudder, to be steered. In winds at levels two through four, "a man-of-war with all sail set and clean full would go in smooth water from one to two knots, three to four knots, and four to five knots." Level six wind was "that in which a well-conditioned man-of-war could just carry, in chase, full and

by single-reefed topsails and top-gallant sails," referring to a
ship chasing an enemy with her sails partially bundled and tied
down, or single-reefed, to reduce the amount of canvas catching
the strong wind. By level ten, more sails were down. At level
eleven, only storm staysails flew—special sails, small and strong,
used not so much to propel a boat as to control it. At level twelve,
Beaufort captured the terror of a hurricane in six words: "That
which no canvas could withstand."

Beaufort was not the first to divide and categorize winds by
their strength. Among the many who came before him was
Tycho Brahe, the rich Dane who some say inspired Shakespeare's *Hamlet*, also known for revolutionizing sixteenth-century
astronomy and for his partially missing nose, a flaw earned in a
duel and hidden by a metal prosthetic. There was also Robert
Hooke, the seventeenth-century scientist credited with the first
use of the word "cell" to describe the appearance of living tis-
sues under the microscope, whose wind scale ranged from one
to four. There was, too, a Dutch scale in the time of Hooke that
ratcheted up from a "dying breath" to a "listless wind" to a
"muzzler" to a "rogue wind." And there was the wind scale of
Anders Celsius, a man remembered today for temperature rather
than wind. At his level four, Celsius reported, the trunk of a
large oak tree in the garden near his famous observatory in
Uppsala, Sweden, "swayed vehemently."

All of these scales, when created, had one thing in common.
They endeavored to create a common language that would allow
scattered observers—most of whom had never seen an instrument capable of measuring wind speed, some of whom did not
even know that such instruments existed—to document the
strength of the wind. At sea, sailors who could not talk reliably
about wind speed in knots or miles per hour could talk mean-
ingfully using Beaufort's scale. Although Beaufort's scale was
neither the first nor the most original of the many scales pro-

posed, it is the one that caught on, and the one that remains with us today despite the fact that anemometers—instruments for measuring wind speed—have become commonplace and measuring wind speed has become routine. Routine, that is, if the wind is not blowing hard enough to rip anemometers away from their bases and send them spinning off into the maw of a raging storm.

Captain John Voss, an adventurous seagoing carpenter and captain, writing in 1913, complained of fickle winds: "During the following two days we accomplished a hundred miles of our course, and were then becalmed for three days." He was off the north coast of Australia at the time, in a boat he had sailed from Vancouver, on the west coast of Canada. The boat was a dugout canoe, thirty-eight feet long, carved from a single log of red cedar in the tradition of the Nuu-chah-nulth people of Canada's west coast and converted by Voss into a sailing yacht. It had no motor. When the wind stopped, Captain Voss stopped.

During his three days of calm north of Australia, his canoe wallowed in swells, the sails empty of wind and slapping incessantly under a burning sun. He spread an awning. To fight the sweltering heat, he poured buckets of seawater over the awning, passing the time in what he considered to be "comparative comfort." Compared with what he did not say.

Later in life, at fifty-four years old, Voss weathered a typhoon off the coast of Japan in a twenty-eight-foot sailboat, a boat smaller but more seaworthy than his dugout canoe. He took the fact that he and his companions survived as proof that a small boat could pass safely through the strongest winds.

"The wind," he wrote, "blew with such force that it was impossible to stand on deck." He counted on a sea anchor—an underwater parachute run out in front of his boat—to hold the bow pointing into the wind, allowing the boat to ride in relative safety. But the wind and waves pulled and jerked and stretched the line to the sea anchor, yanking at it, sending jolts up its length that made the boat shudder. The line, succumbing to the wind's mad abuse, snapped. The boat turned sideways to the wind and flipped, but then righted herself on a breaking wave. Voss was in the sea but somehow made his way back on board. One of his shipmates was in the sea twice, but both times the man regained his place on board.

The boat took on water. Cans of food, bedding, books, and a smashed gramophone floated, awash in the cabin. Voss and his two shipmates broke out buckets and bailed.

They reached the eye of the storm, where the wind died but the seas raged around them. The eye of a typhoon, like the eye of a hurricane, is by all descriptions an odd place, with high seas but still air. The edges of the eye are often described as "walls," as if the raging wind takes on the qualities of a solid structure.

"The barometer," Voss wrote of his time in the eye, "registered twenty-eight twenty-five." That is, he read a barometric pressure of just over twenty-eight inches of mercury, or about 957 millibars. A pressure reading of twenty-eight inches of mercury indicates that the air molecules are standing apart from one another, not densely packed, and that the air molecules from regions of higher pressure, air molecules crowded together and impatient for space, will soon be in motion, eager to race toward regions of lower pressure.

To understand one aspect of the workings of a typhoon or a hurricane, stir tea or coffee in a cup, imitating the circular winds blowing around the center of a hurricane. In the cup, the liquid will rise around the outer circumference and be depressed in the center. But for the spinning, the liquid would flow downhill to the center of the cup, but centrifugal force keeps it piled high around the edge. The storm in the cup mimics the storm in the ocean. In the ocean storm, in the true storm, centrifugal force is balanced by the suction of the low-pressure center. Centrifugal force—the same force felt in spinning-barrel carnival rides, the same force that holds riders to the sides of the spinning drum, really an apparent force rather than a true force— wants to send the winds outward, but the low-pressure center holds them in. The wind spins in what atmospheric scientists call cyclostrophic balance, with centrifugal force seeming to throw air molecules outward and the low-pressure center pulling them inward.

In the eye of the storm for thirty minutes, Voss and his shipmates had neither the time nor the background to understand cyclostrophic balance. They worked frantically on deck, securing what was left of their broken masts, knowing that they would soon smash through a wall of moving air. When the wind came, they went below, secured all the hatches, and waited.

Several days later, reaching Japan, Voss found uprooted trees and a ruined village. "If you feel sure that a typhoon is approaching," he wrote, "prepare to meet it, because it is a tough customer."

*Rocinante's* sails billow. Unless the wind blows from directly behind, sails are not just parachutes catching air. They are more akin to wings. Although details around their surprisingly complicated physics are still debated, sails are often described as airfoils, their curves accelerating the passage of air on the downwind side. When air moves faster across the front of a sail, its molecules spread out and its pressure drops. There are fewer molecules in a cubic foot of air in front of the sail than in a cubic foot of air behind it. The wind in front of the sail creates a void. The wind behind it wants to fill that void. *Rocinante* moves, drawn forward by the wind.

Sails working together—the jib at the front of the boat, the main above my head, and the mizzen just behind me—develop more force than a single sail of their combined sizes. One sail creates an upwash that powers the other.

Sails, being fabric, are not perfect wings. They produce a twisted flow field. There are issues of camber and angle of attack, of stall and luffing. But part of the beauty of sailing is that an attractive sail is, generally speaking, an efficient sail. The most aesthetically pleasing trim of a sail approximates the most ideal physics for that sail. Wrinkles and sags are not only ugly, but inefficient. Perfectly straight lines add neither beauty nor speed.

With my co-captain manning the helm, I walk forward, looking up at *Rocinante*'s sails, watching their gentle curves. I walk aft and tighten the line holding the jib—the jib sheet—and I loosen a line holding the mainsail—the main sheet. The beauty of the curves improves. The speed of the boat increases.

In one form or another, sails have been around for thousands of years, but the final word on sails has yet to be written. The demands of trade and combat at sea drove innovations, but so, too, did sailboat racing, a sport that can be traced to the 1600s in the Netherlands and England. The America's Cup race, the most famous of all races under sail, began in 1851, on the Isle of Wight in England. That first race was won by the schooner *America*, giving the trophy its long-lasting name. The trophy itself, the America's Cup—by some known fondly as the Auld Mug—is a lavishly ornate silver pitcher of no functional value, an antique that might, but for all its history, find its way to a dusty corner. But its history is real and meaningful to those in a position to lust for high-end competition. And that competition yields not only ego boosts for those who may need such things not at all, but also innovations to that ancient wind-catching device, the sail.

In the 2013 America's Cup race, the wing sail took the spotlight. Like all sails, the wing sail has the aerodynamic qualities of a wing, but unlike other sails, the wing sail looks like a wing even at rest. It looks like an airplane wing attached vertically to a mast.

In the 2007 America's Cup race, before the advent of wing sails, racing speeds were around ten knots, or just under twelve miles per hour. Ten knots is — or was — fast for a sailboat. In the 2013 race, with wing sails, boats would reach seventeen knots and then spring out of the water, their hulls lifted on hydrofoils. Once standing on hydrofoils, the boats could approach forty knots.

Boats with wing sails are not bargains. Larry Ellison, cofounder of the software company Oracle and the 2013 America's Cup winner, describes his response to a question about the cost of these boats. "Someone once asked me if it's worth one hundred million dollars to win the America's Cup," he said. His answer: "It's certainly not worth one hundred million dollars to lose the America's Cup."

Here in Galveston Bay, not yet clear of the jetties, close to noon on the first day of our passage, *Rocinante* makes five knots. I whistle softly but with feeling. Ellison's boat might be faster, but *Rocinante* was far less expensive. And she barely leaks at all.

Rear Admiral Beaufort also appears as a historical footnote to one of history's great voyages. Beaufort, at the time acting as the Royal Navy's hydrographer, drew up orders for the voyage of the *Beagle* in 1831. These orders included instructions on recording the wind: "In this register the state of the wind and weather will, of course, be inserted; but some intelligible scale should be assumed, to indicate the force of the former, instead of the ambiguous terms 'fresh,' 'moderate,' &c., in using which no two people agree." The instructions required logbooks to begin with a pasted copy of his scale, on page one, to inform observations made by officers of the watch.

At the request of the *Beagle's* captain, Robert FitzRoy, Beaufort also used his network to identify a young scientist. "Some well-educated and scientific person," Captain FitzRoy wrote to Beaufort, "should be sought for who would willingly share such accommodations as I had to offer." Beaufort found a young scientist named Charles Darwin. Although no one would have guessed it at the time, Captain FitzRoy would eventually become a vice admiral and a pivotal figure in the history of wind.

The breeze picks up. I feel gusts. Captain John Smith called gusts of wind "flaws of wind," and Daniel Defoe referred to gusts as "frets of wind." They had names for gusts, but no understanding. They had no idea why the wind blew. An understanding of why the wind blew—wind at the planetary level, at the level of the globe's great surface rivers of air, at the level of the trade winds—would come later. The planetary level, it turns out, may be the simplest level. An understanding of weather fronts, and the winds brewed when they meet, would come even later. And an understanding of the high-altitude winds that sometimes delay airline flights—jet streams—would come later still.

And it was not until the late 1960s that chaos theory began to explain why Lewis Fry Richardson's belief in numerical forecasting fell short of its promise, why simple observations combined with the laws of motion and mathematics could not predict future winds with an accuracy comparable to that of astronomers with their planetary positions, why something that is fundamentally deterministic is all but impossible to predict with anything resembling accuracy for more than a handful of days. Chaos theory would reveal how tiny differences today could lead to big differences a few days out.

Part of the future of weather forecasting lies in the realization that tiny differences today mean everything for big differences next week and the week after that and the week after that. Part of the future of weather forecasting lies in getting more data now to improve our ability to predict what will come later. And part of the future of weather forecasting lies in understanding and accepting uncertainty, in knowing that the wind will do what the wind will do, and not necessarily what mere humans, armed with physics and mathematics and computers, think it will do. Understanding this, truly appreciating the depth of its meaning, requires an understanding of the discoveries that brought us to this point and an appreciation of the many facets of moving air.

# Chapter 2

# THE FORECAST

For shorthanded crews—sailing couples, crews of two—fatigue characterizes the first few nights of any passage. Taking four-hour watches at the helm, the crew of *Rocinante* grows tired. Even off watch, sleep proves elusive, due in part to the breaking of what passes for normal sleep patterns, the end of eight-hour nighttime rests; in part to the irregular creaking of the vessel, the slapping of her sails, the occasional thud of a wave against her hull; and in part to the worries of the voyage. What has been left behind? Of the many items that we did not repair or replace at the dock, which will we miss most? What will go wrong?

On the morning of our third day off the dock, the light of the rising sun, more or less in front of us, douses fatigue and anxiety. The light of the rising sun revitalizes.

For now, storms are not a real concern. We sail on the tail end of a norther. In this part of the world, at this time of year, northers tend to restore equilibrium. They sweep down through Montana and Wyoming carrying the air of Edmonton and Calgary and Saskatoon. Although they are easy to forecast as they move across the nation, from one weather station to the next, the forecast does nothing to diminish their force or their sudden cold, their forty-knot winds or their twenty-degree temperature drops. But the beauty of a norther for a boat heading east under sail, for a boat making its way from Galveston to the Yucatán, is that it comes from the north.

This particular norther came on with considerable strength. More-experienced crews might have sailed at the height of the norther's anger, with sails reefed, going east as fast and as far as possible, taking advantage of every minute of the north wind. Before we left the dock three days ago, the norther had already softened. Now, for a moment, its winds die altogether. After five minutes, they return as gusts of twenty knots, puff for a minute or two, and then clock around to the northeast, losing strength.

With the norther passing, pressure differences across the region diminish. Meteorologists, with an insatiable appetite for maps and mapping, draw lines to show pressure changes, as though the atmosphere itself is a tangible landscape whose contours can be plotted on a topographic map. Lines joining points of equal pressure are called isobars. On a topographic map, widely spaced lines show coastal plains and small hills, while closely spaced lines show the Appalachian Mountains, the Brooks Range, the Rockies. On the weather map, widely spaced lines show areas of slowly changing pressure, while closely spaced lines show areas of rapidly changing pressure. With slowly changing pressure, there is little wind. With rapidly changing pressure, with closely spaced isobars, the wind will blow as if roaring down the face of a mountain.

On this particular day, a meteorologist might note that the isobars on the weather map for the northern Gulf of Mexico have spread apart, that today's weather map, with fewer lines, requires less ink than yesterday's weather map. The meteorologist, in seeing this, would know that the wind is dying. Looking at the spacing of the bars, the meteorologist could estimate wind speeds.

Aboard *Rocinante*, through gloves and hats and foul-weather coats, we feel a breath of warm air. An easterly?

We do what we can to make easting. As the wind changes, we change course. When the wind zigs we zag, and when the wind zags we zig. We leave behind a serpentine wake.

The Mississippi River discharges somewhere off the port side, beyond the horizon. We sail well clear of the coast, struggling to reach a point south of Pensacola, Florida, or at least south of Mobile, Alabama, before turning to the Yucatán and Isla Mujeres.

It is sometimes said that Edmond Halley drew the first weather map. If by weather map, one means the synoptic weather map of newspapers and television forecasts, he did not. What he drew was entirely different. In 1686, he drew a map of the trade winds based on a collection of accounts from sailors and from his own observations. "To help the conception of the reader," he wrote, "in a matter of so much difficulty, I believed it necessary to adjoyn a Scheme, shewing at one view all the various Tracts and Courses of these Winds; whereby 'tis possible the thing may be better understood, than by any verbal description whatsoever." In other words, he wanted a simple representation of wind patterns, not on any one day but in general terms. What Halley drew was not the kind of weather map seen in newspapers and on television, but an early example of what became known as a thematic map. Halley's map showed wind trends, but it did not show a snapshot of actual conditions.

Benjamin Franklin, too, was interested in trade winds, but his contribution came not from his interest in the trades but from his interest in storms. That interest can be dated to at least 1734, but it culminated in the storm of October 21, 1743, the storm that ruined Franklin's observations of a lunar eclipse.

The wind that day, at Franklin's location in Philadelphia, blew from the northeast. Franklin assumed that his friends in Boston, upwind, had experienced the same storm before him. He thought they might have seen the eclipse after the storm passed. But then he came across the October 24 issue of the

*Boston Evening Post.* "Last Friday night," it said, "soon after a total and visible eclipse of the Moon (which began about nine and ended past one o'clock) came on a storm of wind and rain, which continued all the following day with great violence." The wind Franklin had felt was blowing toward Philadelphia from Boston, but the storm had traveled away from Philadelphia, toward Boston. The storm, from where Franklin sat, had blown upwind.

After six years of rumination, he wrote to a friend in Connecticut: "Though the course of the wind is from northeast to southwest, yet the course of the storm is from southwest to northeast; the air is in violent motion in Virginia before it moves in Connecticut and in Connecticut before it moves at Cape Sable."

Franklin also published his thoughts in the form of a note on what was possibly the first map of Pennsylvania published in America. The map, drawn by a surveyor named Lewis Evans, was printed by Franklin's print shop in 1749. As some maps are, the Evans map is a work of rare beauty. In the lower right corner, in space claimed by the Atlantic Ocean, two square-rigged ships sail. Rivers snake across the parchment. Regularly spaced scratches mark mountains. Delaware Bay, the estuary of the Delaware River, sweeps northwest from Cape May.

Elsewhere, in areas that would otherwise be left blank owing to an absence of topographical details, extensive notes in flowing script inform users about the region. Near the top, one note says, "The Country between this and St. Lawrence is full of Mountains, Swamps, and drownded land." The Hudson River is labeled "Hudson's or North R." Just east of the Hudson River, a note says, "The Bounds of N. York and Massachusetts Prov. Are not Settled." An entire paragraph provides "Remarks on the Endless Mountains." And this, in the lower left: "I am afraid the Longitude is not so exact as could be wisht."

This map was in no way a weather map, but in the upper left corner, on the shores and shallows of Lake Ontario, a paragraph of several hundred words describes regional climate, beginning with "the Extreams of the Barometer" and the temperature range based on "Fahrenheit's Pocket Thermometer." In this paragraph, buried in a single sentence, is Franklin's realization: "All our great Storms begin to Leeward: thus a NE Storm shall be a day sooner in Virginia than in Boston." The storms, Franklin saw, did not come from the same direction as the wind that he experienced on the ground. The wind that Franklin felt in 1743 blew hard from the northeast. The storms somehow originated on the downwind side — in Franklin's words, the leeward side, to the southwest. To a man on the ground or to a ship on the water, the storms moved against the wind. Anyone trying to forecast weather would need to know what was happening not only on the ground but also aloft.

Playing with *Rocinante's* VHF radio, I hear a crackling word or half a sentence of a static-ridden weather forecast, possibly the forecast for the offshore waters from twenty to sixty nautical miles out, from Pascagoula, Mississippi, to Pensacola, Florida. We should be well out of range, but there must be a repeater somewhere out here, or strange atmospherics, allowing a few words to sneak through, an electronic voice saying something like "seas to three feet" and "winds ten to fifteen knots," repeating the message again and again, never tiring, except to the listener. But no matter how long I listen, the static outperforms the voice, and though I am left without details, I am comforted by what seems to be the absence of storm warnings.

We see, occasionally, oil and gas platforms, yellow or orange or gray or black. We have more than one thousand feet of water under our keel. At this depth, few platforms stand on legs. They are hulking functional beasts, entirely deprived of beauty, looking nothing like ships even though they do in fact float. They are also weather stations, transmitting information shoreward, part of the network that pumps data into the ever-changing synoptic weather maps, a network that fills the cells of something that grew from Lewis Fry Richardson's dreamed-of forecasting grid.

All forecasting, numerical or otherwise, starts in the same place. It starts in the here and now. Someone has to open a window to see what the world is doing.

Ideally, that someone is armed with a few instruments. Ideally, that someone is in communication with like-minded someones at other locations, all armed with similar instruments. Not one or two or three like-minded someones, but thousands. Together, pooling their measurements of temperature, atmospheric pressure, and humidity and their observations of cloud cover and wind speed, they begin to understand the weather. They establish the foothold needed to generate a forecast.

In a time accustomed to forecasting, when televisions, radios, computers, and smartphones offer forecasts in reasonably accurate hourly increments, the barriers to the early forecasters are difficult to fathom. But for perspective, consider the reaction of the British House of Commons on June 30, 1854, to the words of the member for Carlow, Mr. John Ball, two decades before the first newspaper weather map was printed. During discussions regarding the establishment of a meteorological office, the function of which would be to collect weather statistics, Mr. Ball had the nerve to suggest the possibility of forecasting. As statistics piled up, he thought, they would provide insights not only about the past but also about the future. "In a few years," he said to the House of Commons, "notwithstanding the variable climate of this country, we might know in this metropolis the condition of the weather twenty-four hours beforehand."

The reaction, according to the official record: laughter.

In referring to "great storms" in 1749, Benjamin Franklin meant the kinds of storms now known as extratropical or midlatitude cyclones. Even Franklin could not anticipate the nomenclature of weather that would evolve well after his death, but he saw that a strong wind blowing from the northeast arrived from the southwest.

Northers, though, blow from the north, and they travel, more or less, from the north to the south. Our norther, by midafternoon on our third day at sea, is almost blown out. My co-captain pinches, meaning that she points *Rocinante* as far upwind as she will go before the sails are backwinded, before they collapse and we stop making headway. This tack—this heading relative to the wind's direction—heels *Rocinante* over, slanting her decks. My co-captain sits behind the helm in an awkward leaning posture, legs braced against the side of the cockpit. We sail in the most picturesque of tacks, the classic leaning pose of a boat tilted by the wind. But beauty comes with discomfort and inefficiency and strain on the rigging. Beauty of this sort comes with books falling off shelves and teakettles sliding to the rail that surrounds *Rocinante*'s two-burner stove. Most of the wind's energy pushes the boat sideways and pins her at an angle rather than pushing her forward. Sailing upwind—sailing as close to upwind as we can—would not be our first choice. But we have to make easting, so we have no choice.

And even pinching, we cannot go due east. At best we are headed southeast. My co-captain has begun what we hope will be a shallow arc toward southwest Florida that will put us well east of the Yucatán before we are forced to turn due south. If we wander too far south before clearing the tip of the Yucatán, we will regret it later, when we will have to fight both the wind from the east and the Yucatán Current flowing along the coast. We know all this only from books. We would rather not learn experientially of the difficulties of sailing against the Yucatán Current. And so for now my co-captain does her best to press eastward, her long hair blowing free, her eyes tired but shining blue, content.

We are, by afternoon, one hundred miles offshore. On the VHF radio, I hear nothing but static. The sun warms my face. I shed my foul-weather jacket. I shed my foul-weather pants. I relieve my co-captain at the helm so that she can catch a few hours of sleep. She goes below, and through the open companionway hatch I watch her dozing, her hair loose and tumbling across her face.

*Rocinante* makes five knots through the water.

In the 1800s, in a world without airplanes, people traveled by ship. In a world without forecasts, ships sank in storms with surprising frequency. Even with steam-powered steel ships, even close to the refuge of protected harbors, disaster struck.

Two disasters stand out in the history of weather forecasting. The first was in 1854, the same year the British House of Commons laughed at the prospect of forecasting and also one year into the Crimean War. British, French, and Turkish ships were anchored off the cliffs of Sevastopol, in the Black Sea, and there they faced a grand assault. The assault came not from enemy troops but from nature itself. Raging winds added to the already abundant horrors of what is sometimes called the first modern war.

The *Rip Van Winkle,* a sailing transport, broke away from her anchors and was carried broadside against the cliffs. Witnesses later reported that she was smashed to pieces on impact. There were no survivors. The *Progress* and the *Wild Wave* followed the *Rip Van Winkle's* lead, and then the *Kenilworth*. And the *Wanderer*. The *Prince*, a British supply ship carrying winter gear and medical supplies, once described as "the pride of the transport service," resisted, running her steam engine to ease the strain on her anchors. But she, too, was eventually thrown onto the rocks. Of the 150 men on board, 6 survived.

Up the coast, beyond the cliffs, the wind carried a French steam frigate onto the beach. And most famously of all, the French battleship *Henri IV,* a three-decker carrying one hundred guns, snapped four anchor chains and blew toward shore, grounding on sand where she survived the night. The ship's chaplain, Abbé Bertrand, described furniture flung around within the ship and the horribly sudden snapping of the last two anchor chains. "The sea was furious," he wrote, "and bellowed so as to prevent us from hearing each other." Later, the frigate's remains would be used to construct a fortress.

From the log of the *Samson,* anchored off the Katcha River: "Extra allowance of spirits issued." From the log of the *Bellerophon,* also anchored off the Katcha: "ship struck by other ships" and "bowsprit bulwark carried away."

By morning, the shore was littered with wrecks. More from the *Bellerophon's* log: "Daylight; observe a Turkish battleship on shore; Onshore 5 English and 5 French transports, *Henri IV* and 2 French steamers."

Back in France, the minister of war requested the help of Urbain Le Verrier, the man who discovered the planet Neptune. Le Verrier found Neptune not with a telescope but with his knowledge of oddities in the orbit of Uranus. If he could predict the existence of an unseen planet, surely he could predict changes in the weather here on earth.

After collecting more than 250 weather reports from across

Europe for November 10 through November 16, 1854, Le Verrier concluded that the path of the storm could have been predicted. The French navy could have been warned. The loss of the battleship *Henri IV* could have been prevented.

On Le Verrier's advice, Louis-Napoléon Bonaparte, nephew and heir of Napoléon Bonaparte, called for the establishment of a storm warning system. That system would rely on a new technology, the electric telegraph.

What we think of today as the telegraph had made its debut just ten years earlier. In 1844, Samuel Morse, sitting in a chamber of the Supreme Court in Washington, DC, sent Albert Vail, sitting in a railroad depot in Baltimore, a simple message: "What hath God wrought?" One possible response, never sent but nevertheless viable: "God hath wrought an instrument that will facilitate weather forecasting."

Four years after his famous first transmission, Morse wrote a letter to his brother. It was an ordinary letter, on paper. In it, and in other documents, Morse defended his position in commerce and history as the inventor of the telegraph. "I have been so constantly under the necessity of watching the movements of the most unprincipled set of pirates I have ever known," he wrote, "that all my time has been occupied in defense, in putting evidence into something like legal shape that I am the inventor of the Electro-Magnetic Telegraph!!" A newspaper article described accusations that he had stolen the invention from others, that it had originated in Germany.

Morse's defensiveness was not without cause. He had not in fact invented the telegraph from scratch. It had grown out of primitive visual signals—flags hoisted onto poles or wooden shapes raised on wooden arms extended from towers. By the late eighteenth century, well before Morse's efforts bore fruit, messages were being transmitted across France and England and even Ireland, where a young Francis Beaufort—at the time not yet addressed as Rear Admiral or Sir—was among those employed to construct a network of optical telegraphs. But these visual telegraphs were slow and uncertain. Messages were relayed from one station to the next to the next. Fog and rain and snow sometimes obscured signals. By the time a message reached its intended destination, it could be garbled to the point of unintelligibility.

As early as 1753, a suggestion had been made for an electric telegraph that would rely on twenty-six insulated wires, one for each letter of the alphabet. A pith ball, attracted by electricity, would move in response to currents at the receiving end. Later, in electrolytic telegraphs, current was used to liberate hydrogen, and bubbles were seen at the receiving end. By 1830, a system sent messages eight miles across Long Island, changing the acidity of a liquid at the receiving end, with the messages appearing on a moving strip of litmus paper. In 1831, Joseph Henry, who would later be appointed the first director of the Smithsonian Institution, developed a telegraph that relied on electromagnets. Current swung a hanging magnetized steel bar into a bell.

Morse did not intend his famous 1844 message to be received as a series of clicking dots and dashes, but as a series of scrawls electrically transcribed onto scrolling paper. The scrawls looked something like Ws and Vs. Those on the receiving end heard the receiver clicking. They realized that paper might be unnecessary. Signals could be received audibly, as clicks and pauses. Morse and Vail, with others contributing, developed what would become known as Morse code.

What had been possible before through the use of flags or other signaling devices was rendered convenient by the electric telegraph. A good telegrapher could transmit or receive thirty words each minute. Information could be shared without delays, and with that information storms could be tracked. By 1859— just five years after the storm that beached and destroyed the

*Henri IV*—Le Verrier's network was distributing daily bulletins on current weather recorded at nineteen stations.

But claiming that the 1854 storm could have been predicted and routinely predicting future storms were two different matters. Only eight years earlier, François Arago, Le Verrier's highly respected predecessor at the Paris Observatory, had written, "Never has a line published with my consent, authorized any one to imagine it to be my opinion that it is possible, in the present state of our knowledge, to announce with any degree of certainty, what weather it will be a year, a month, a week, I shall even add, a single day, in advance." And this: "Whatever may be the progress of the sciences, never will observers who are trustworthy, and careful of their reputations, venture to foretell the state of the weather."

Arago's caution was well founded. Then as now, but even more so then than now, future weather mocked forecasts and shamed forecasters, including Le Verrier. Winds that should have blown rested. Winds that should have rested blew. Rain expected to fall emerged as clear blue skies. Clouds came in, entirely uninvited.

When the French minister of war openly criticized Le Verrier's forecasts in 1863, Le Verrier likely thought of his predecessor's comments about "observers who are trustworthy." In 1867, Le Verrier did what he could to quash the nascent forecasting service that had grown within the Paris Observatory. In what appears to have been an effort to protect his reputation and position, and to retain his status as a trustworthy observer, he disassembled the very forecasting capabilities that he had once nurtured.

The 1854 storm that wrecked the *Prince* and the *Henri IV* drew the attention of the British just as it drew the attention of the French. In its wake, the British Board of Trade established the

Meteorological Office, later to be known as the Met Office, the national weather service of the United Kingdom. While considering potential candidates to head the new office, the Board of Trade consulted Robert FitzRoy, the man who had commanded the *Beagle*, the ship that had carried Darwin around the globe. "A captain of considerable length of service should be selected," FitzRoy wrote, "an actively disposed officer living in or near London." He noted that the role would require a person willing to work long hours and a person with skills beyond those of a typical ship captain. The job would require a person who understood winds and currents and temperature and magnetism, "an Arago—a Whewell—a Rennell—a Reid—a Sabine—and a Faraday and a Herschel combined under a Humboldt," he noted, providing a who's who of the sciences of the day and leaving the impression that no one would be good enough for the job.

FitzRoy himself turned out to be the best choice for the task at hand. His résumé was impressive. As a young man, he had studied Newton's work and mathematics, as well as navigation and the more gentlemanly arts of fencing, dancing, and painting. When he sailed, he sailed with a library. As a young lieutenant, he packed four hundred volumes—books on the science of the day, but also books that helped him with his Latin, Greek, French, Italian, and Spanish. Like all captains, he had experienced weather, good and bad, including a gale that ripped across the Argentine Pampas, a wind known as a pampero that appeared out of nowhere to very nearly capsize his first ship, killing two of his seamen. Throughout his time at sea, he kept notes on various observations—not just passing notes but systematic notes, recording his world in the manner of a person interested in a greater understanding, in the manner of a person who saw himself as a scientist. The notes, of course, included wind directions and strengths, as well as barometric pressures and temperatures. And his writing sometimes portrayed his sense of duty and ambition. "Those who never run any risk," he wrote, "who only sail when the wind is fair; who heave to when approaching land, though perhaps a day's sail distant; and who even delay the performance of urgent duties until they can be done easily and quite safely; are, doubtless, extremely prudent persons; but rather unlike those officers whose names will never be forgotten while England has a navy."

*Robert FitzRoy, near the time of his first weather forecast. (Image from Wikimedia Commons)*

After commanding the cruise that made Darwin famous, but well before the Board of Trade considered meteorology, FitzRoy published, with Darwin, a four-volume account of their voyage,

*Narrative of the Surveying Voyages of His Majesty's Ships* Adventure *and* Beagle. Darwin's own *Journal and Remarks, 1832–1836*, later sold alone under the title *The Voyage of the* Beagle, made up the third volume. During preparation of the book, the two men corresponded. They were, at the time, clearly friends. They discussed, among other things, the inclusion of advertisements in their book. "I am sure you will agree with me in thinking it desirable to avoid swelling the volumes with ordinary advertisements," FitzRoy wrote in 1839, just before the book came out. He went on to say that including advertisements for Darwin's other books was fine by him, that his objections applied only to ordinary advertisements. "I have been intending to call on you these many days but have not had time for the excursion," he added.

After the book came out, FitzRoy was appointed to govern New Zealand, an ill-fated assignment that ended in a controversy over the rights of natives. Over time, his own remarkable achievements aboard the *Beagle* were overshadowed by the increasing fame of his onetime ship's naturalist. By 1850, for all practical purposes, FitzRoy had been put out to pasture. In 1854, when he was appointed Meteorological Statist to the Board of Trade, with a staff of three, he was called out of retirement. He threw himself into the job.

The job, to be clear, was that of a record keeper. He was tasked with keeping weather statistics, not with forecasting future weather. Thanks to the 1854 storm that wrecked the *Prince* and the *Rip Van Winkle* and the *Progress* and the *Henri IV*, FitzRoy became a statist.

In 1859, five years after FitzRoy started his job as Britain's chief weatherman, five years after the disastrous storm on the Crimean coast, a new maritime disaster shook the world. A ship called the *Royal Charter* sailed from Australia bound for Liverpool. She was powered by both sails and a steam engine, and she carried passengers as well as riches from the gold rush in Victoria. After fifty-nine days at sea, she stopped briefly in Ireland. Her captain, mindful of his ship's reputation for speed, was eager to reach Liverpool. He set sail for what should have been an easy daylong crossing, but he soon noted a change in the wind. What had been a favorable wind now came on his bow. He lowered his sails, relying on his steam engine to bash through what was clearly a growing storm. The sky turned gray as Wales came into view.

The *Royal Charter* rounded Anglesey and was abruptly confronted with the most unfavorable of changes. The wind, which had been blowing from the east, now shifted to the north and increased in strength. Blowing from its new direction, it reached level twelve on the Beaufort scale. It was air moving at more than one hundred miles per hour, and it was pushing the *Royal Charter* toward shore.

*The wreck of the* Royal Charter *inspired FitzRoy to move from keeping the statistics of past weather to forecasting future weather. The painting, though dramatic, is not consistent with accounts from survivors. (Image from Wikimedia Commons)*

In the open sea, level twelve winds generate waves more than forty feet tall. They turn the sea white with foam. They strip away the tops of waves, forcing crew and captain to peer through driving spray. But peer as they will, sailors are powerless in winds this strong. Men can fight the helm to maintain a misplaced sense of control or they can cower in the hold, in both cases knowing that their actions will no longer impact their destinies. They live by the mercy of the wind or die by its ambivalence. Level twelve winds are hurricane-strength winds.

The captain watched as the wind blew his ship closer to the rocky coast. He lowered two anchors. Progress toward the rocks slowed. He launched rockets, signaling that his was now a ship in distress, a ship needing assistance, but in winds of such strength there could be no assistance aside from prayer.

The ship strained hard against her two anchors. The captain ordered his men to cut away the masts to reduce windage.

After two hours, one of his ship's anchor cables snapped. After another hour, the second anchor cable snapped. The ship and all she carried moved toward the shore. The bottom of the ship's hull made contact with the bottom of the ocean.

By virtue of short-lived luck, the ship touched down on sand within a stone's throw of the rocky shore. A seaman dove into the raging waves and swam toward land, carrying with him a line to the ship. Miraculously, he reached land. People who lived nearby were there to help. As they rigged a line to begin bringing passengers ashore, a wave broke across the ship's deck, carrying away a group of women and children.

The tide rose. The ship herself, her moment of luck expired, floated loose from the sandbar and was blown onto rocks. She broke in two.

More than four hundred men, women, and children died. Many were dashed against the rocks, pulverized by the pounding waves. Others sank to their deaths, weighed down by belts stuffed with the gold they had found in Australia. The captain, one survivor reported, was last seen in the water, struggling next to the hull, when something fell from the ship and struck him in the head.

Charles Dickens wrote of the sinking: "So tremendous had the force of the sea been when it broke the ship, that it had beaten one great ingot of gold deep into a strong and heavy piece of her solid iron-work, in which also several loose sovereigns that the ingot had swept in before it had been found, as firmly embedded as though the iron had been liquid when they were forced there."

Only thirty-nine of those aboard the *Royal Charter* survived. The bodies of the dead continued to wash ashore for weeks after the ship was lost.

The sun sets off the starboard quarter. In plain English, we remain headed more or less southeast. But every few minutes, I have to turn to the south to keep wind in the sails, putting the sunset directly to starboard, and a few minutes later, with the shifting wind, I turn back.

In the early twilight, I break out a computer and open saved files and screen captures from PassageWeather, a free sailing weather website. From a screen capture explaining the site: "The majority of our forecasts are derived from the 0.5 degree GFS weather model from NOAA/NCEP." In other words, they use the Global Forecast System model produced by the National Centers for Environmental Prediction, part of the US government's National Oceanic and Atmospheric Administration. This is the sort of tool that appeared in the dreams of Lewis Fry Richardson. Just as he suggested, the weather models of today break the world up into a grid, with each square on that grid forming a cell, and with mathematics applied to each cell and the interactions between adjacent cells.

More from PassageWeather: "For North America, we create higher-resolution Surface Wind charts using data from the 12 km NAM model from NOAA/NCEP." Decoded, this means that the site uses a better model for North America—the North American Mesoscale Forecast System, a model with higher resolution and smaller cells.

My saved files show a seven-day forecast downloaded three days ago, just before we left the dock. One set of files shows wind direction and strength. On a map of the Gulf of Mexico, dozens of tiny lines—properly called wind barbs—show wind direction. At the tail end of each of these lines, diagonal strokes show wind speed—a half stroke for five knots, a full stroke for ten knots, one and a half strokes for fifteen knots. At fifty knots, strokes are replaced by the classic mariner's long triangular flag. I hope never to sail across strokes replaced by a flag.

For the moment, at *Rocinante*'s current location, adjacent wind barbs point southeast and east, one with a half stroke, most with one stroke, a few with one and a half strokes. No flags fly—not a single flag in all of the Gulf of Mexico.

I scroll through the maps for the next four days—eight maps, one issued for every twelve hours. The forecast drawn on the screen has merged, in my mind, with reality. It is a reality that has yet to come.

What matters most to me are not the details of the models, not whether it is the NAM model or the GFS model or some hybrid, but the authoritative lines on my screen. Authoritative wind barbs on a screen offer a concrete future, or at least the illusion of one.

While looking at the barbs indicating the speeds of winds-to-be, I fail to pay attention to the helm. *Rocinante* turns upwind. With pressure off the sails, she levels out. Her forward progress rapidly diminishes. Her foresail slaps.

Before she loses all momentum, I turn the helm. Her sails

refill. She heels over, her deck angling beneath my feet. She picks up speed.

Below, my co-captain stirs. She calls out, asking if all is okay.

"Nothing to worry about," I answer, and in the shadows through the open hatch, I see her roll over on the once again slanted berth.

*Rocinante* makes five knots through the water, but not quite in the right direction. With this wind, the right direction is no longer an option, but I take solace in the fact that we are going more or less toward our destination. And I resolve—for neither the first nor the last time—to pay more attention to what I am doing.

To men like FitzRoy, the storm that sank the *Royal Charter* drove home the importance of storm warnings. The storm took not only the *Royal Charter* but an additional 132 ships. The loss of ships and the deaths of passengers, highly publicized across England, could not be ignored.

FitzRoy, though tasked with maintaining weather records, with building a statistical history of rain and wind, had come to the job predisposed to do more. From his voyages around Cape Horn and his surveys of windswept Tierra del Fuego, he was attuned to the weather. The Royal Charter Storm, as it is known to history, pushed FitzRoy to step beyond the bounds of mere record keeping. Past winds had irrevocably affected lives, but the past was the past. The future could also affect lives. Knowledge of future winds could save ships, sailors, and passengers. Although he was first and foremost a mariner, he also saw himself as a man of science. It was time, FitzRoy saw, to apply science to future winds. It was time to offer forecasts.

Like Le Verrier, FitzRoy recognized the telegraph's game-changing potential. By virtue of the invention, FitzRoy had weather information from stations scattered around the nation at his fingertips, and he believed he could use this information to predict tomorrow's weather today. If telegraphed reports tracked a storm headed toward London, the residents of London could be told of the storm before it arrived. FitzRoy coined the term "forecast" and published his first daily weather forecast in the *Times* of London on August 1, 1861.

## THE WEATHER.

### METEOROLOGICAL REPORTS.

| Wednesday, July 31, 8 to 9 a.m. | B. | E. | M. | D. | F. | C. | I. | S. |
|---|---|---|---|---|---|---|---|---|
| Nairn.. | 29·54 | 57 | 56 | W.S.W. | 6 | 9 | o. | 3 |
| Aberdeen .. | 29·60 | 59 | 54 | S.S.W. | 5 | 1 | b. | 3 |
| Leith .. | 29·70 | 61 | 55 | W. | 3 | 5 | c. | 2 |
| Berwick .. | 29·69 | 59 | 55 | W.S.W. | 4 | 4 | c. | 2 |
| Ardrossan . | 29·73 | 57 | 55 | W. | 5 | 4 | c. | 5 |
| Portrush .. | 29·72 | 57 | 54 | S.W. | 2 | 2 | b. | 2 |
| Shields .. | 29·80 | 59 | 54 | W.S.W. | 4 | 5 | o. | 3 |
| Galway .. | 29·83 | 65 | 62 | W. | 5 | 4 | c. | 4 |
| Scarborough .. | 29·86 | 59 | 56 | W. | 3 | 6 | c. | 2 |
| Liverpool .. | 29·91 | 61 | 56 | S.W. | 2 | 8 | c. | 2 |
| Valentia .. | 29·87 | 62 | 60 | S.W. | 2 | 5 | o. | 3 |
| Queenstown .. | 29·88 | 61 | 59 | W. | 3 | 5 | c. | 2 |
| Yarmouth.. | 30·05 | 61 | 59 | W. | 5 | 2 | c. | 3 |
| London .. | 30·02 | 62 | 56 | S.W. | 3 | 2 | b. | — |
| Dover.. | 30·04 | 70 | 61 | S.W. | 3 | 7 | o. | 2 |
| Portsmouth .. | 30·01 | 61 | 59 | W. | 3 | 6 | o. | 2 |
| Portland .. | 30·03 | 63 | 59 | S.W. | 3 | 2 | c. | 3 |
| Plymouth.. | 30·00 | 62 | 59 | W. | 5 | 1 | b. | 4 |
| Penzance .. | 30·04 | 61 | 60 | S.W. | 2 | 6 | c. | 3 |
| Copenhagen .. | 29·94 | 64 | — | W.S.W. | 2 | 6 | c. | 3 |
| Helder .. | 29·99 | 63 | — | W.S.W. | 6 | 5 | c. | 3 |
| Brest .. | 30·09 | 60 | — | S.W. | 2 | 6 | c. | 5 |
| Bayonne .. | 30·13 | 63 | — | — | — | 9 | m. | 5 |
| Lisbon .. | 30·13 | 70 | — | N.N.W. | 4 | 3 | b. | 2 |

*General* weather probable during next two days in the—
North—Moderate westerly wind ; fine.
West—Moderate south-westerly ; fine.
South—Fresh westerly ; fine.

Explanation.
B. Barometer, corrected and reduced to 32° at mean sea level ; each 10 feet of vertical rise causing about one-hundredth of an inch diminution, and each 10° above 32° causing nearly three-hundredths increase. E. Exposed thermometer in shade. M. Moistened bulb (for evaporation and dew-point). D. Direction of wind (true—two points *left* of magnetic). F. Force (1 to 12—estimated). C. Cloud (1 to 9). I. Initials :—b., blue sky ; c., clouds (detached); f., fog ; h., hail ; l., lightning ; m., misty (hazy) ; o., overcast (dull) ; r., rain ; s., snow ; t., thunder. S. Sea disturbance (1 to 9).

*The first published weather forecast appeared in four short lines below the daily weather report on August 1, 1861, in the* Times *of London. This forecast has been reproduced many times since 1861, including on the website of the British Met Office, where FitzRoy once worked.*

That first newspaper forecast, the first newspaper weather forecast that the world ever saw, appeared with little fanfare. The *Times* had been publishing descriptions of the previous day's weather for years, and this first forecast, this historically important forecast, appeared below a table summarizing the previous day's weather. It ran a mere twenty-three words:

General *weather probable during next two days in the —*
North *— Moderate westerly wind; fine.*
West *— Moderate south-westerly; fine.*
South *— Fresh westerly; fine.*

Despite the modest nature of the first forecast, despite the absence of pomp that accompanied its release, FitzRoy knew he was onto something important. From that day forward, for what was left of his life, every working day involved the preparation of a forecast. "Prophecies and predictions they are not," he wrote, not wishing to be lumped with astrologers; "the term forecast is strictly applicable to such an opinion as is the result of scientific combination and calculation."

He worked frantically, trying to make sense of the weather, trying to understand the atmosphere, collecting and collating information delivered via telegraph and collecting and collating ideas developed through letters and discussions with friends and colleagues. But he was a man of what some called practical science, not philosophic science. He looked for patterns and relationships but did not concern himself too much with the underlying causes of what he observed. Via his ever-growing collection of past storm tracks, he learned of trends. If a particular set of conditions existed today, another particular set of conditions usually existed tomorrow. And if his telegraphic network tracked a storm heading north, that storm would probably continue heading north. His intuition guided his interpretation. His approach, though complex by the standards of the times, was simplistic and naive by the standards of modern atmospheric scientists. It ignored the complexities of how and why the atmosphere behaves as it does, and in ignoring the complexities, the approach often produced inaccurate results.

Around him, others—including very influential others—were concerned about the how and why. The science establishment at the time increasingly believed that true advances required the underlying theories and laws of philosophic science.

FitzRoy had, before his first newspaper forecast, attended the annual meeting of the British Association for the Advancement of Science. He was one of more than a thousand participants. Science was, at this time, becoming a recognized profession. It was also confronting conventional religion more openly than it had in the past. And this was the first meeting of the society since the publication of Darwin's *On the Origin of Species*.

In the crowd were two men whose words would become legendary in the world of science. One was Thomas Huxley, the man whom history has dubbed "Darwin's Bulldog," and the other was the bishop of Oxford, Samuel Wilberforce. Darwin himself, during this period of his life, was somewhat reclusive, defending himself and his ideas about natural selection through the post rather than in person, but his friend Huxley, as well as others, planned to defend the merits of evolution by natural selection. They were prepared to argue openly about evolution with Wilberforce, a man of known oratory skill.

During the meeting, FitzRoy presented a paper on British storms to what must have been a yawning and impatient audience, but he stayed on to hear other talks. There was talk after talk, covering topics as varied as sleep, volcanic eruptions, the arrangement of ancient Irish villages, and submarine cables. But history remembers the meeting for the exchange that occurred after a now mostly forgotten professor presented a work titled "The Intellectual Development of Europe, Considered with Reference to the Views of Mr. Darwin and Others, That the Progression of Organisms Is Determined by Law." It was this paper that set the stage for the famous exchange between Huxley and the bishop.

The exchange turned rabid quickly. In an effort to discredit natural selection, Wilberforce inquired about the apes in Huxley's family tree. Although Huxley's exact response was not recorded, it has been paraphrased and set in stone over time. "I should feel it no shame to have risen from such an origin," Huxley is supposed to have said, "but I should feel it a shame to have sprung from one who prostituted the gifts of culture and eloquence to the service of prejudice and falsehood."

Less famously, FitzRoy stood up. He would have been recognized as the onetime commander of the *Beagle* and the man who had spoken rather tediously about storms. Many in the audience knew that FitzRoy and Darwin were long-term if sometimes strained friends and that they had published together.

FitzRoy told the assemblage that he did not believe that Darwin's work on natural selection should have seen the light of day. He did not believe that Darwin offered a logical arrangement of the facts. He said that he "had often expostulated with his old comrade of the *Beagle* for entertaining views which were contradictory to the First Chapter of *Genesis*."

According to some accounts, FitzRoy held a Bible as he spoke.

As far as the biological world was concerned, FitzRoy landed on the side of religion. The Victorian science establishment disagreed.

When FitzRoy began publishing forecasts a year later, the Victorian men of science—and they were mostly men—held that forecasting was, at best, premature. FitzRoy relied too heavily on intuition and rote experience. He had no real explanation for the workings of the weather, no sensible theoretical under-pinning. He might claim that his forecasts were neither prophe-cies nor predictions, that they were based on science, but others did not agree. To the Victorian science establishment, FitzRoy was no scientist.

To some degree, FitzRoy set himself up for criticism. The men of science embraced thinking based on empirical observations in support of theories, not thinking based on the first chapter or any other chapter of Genesis, not thinking based on the intu-itions that grew from life at sea. They wanted methods that could be written down and applied by all objectively, and preferably expressed as equations. But FitzRoy's critics came from other quarters, too. Among the most devastating was one who espoused forecasting based on the position of the moon, another motivated by retaliation, and a third motivated by business interests.

The first man was the naval officer and naval engineering instructor Stephen Martin Saxby. In 1862, Saxby published a book on weather forecasting. On its surface, the book might have bolstered FitzRoy's belief in both the usefulness and the possibil-ity of forecasting. It might have seemed that FitzRoy had found a sympathetic friend, a fellow believer. But the book's content destroyed all hope of fellowship. Saxby believed weather to be controlled by the moon. His book, called *Foretelling Weather: Being a Description of a Newly-Discovered Lunar Weather-System,* was not the sort of work that should be blindly accepted in a soci-ety increasingly influenced by science, and yet a public accus-tomed to almanacs and hungry for a sense of certainty about the future accepted it. An undiscerning public embraced Saxby's work as much as it embraced the work of FitzRoy himself.

Astrological meteorology had a long history. Ptolemy was among the believers. Johannes Kepler, the German mathematician and astronomer known for his laws of planetary motion, was also a believer. In 1686, a somewhat obscure figure named John Goad reinvigorated the old beliefs with his book, *Astro-Meteorologica; or, Aphorisms and Discourses of the Bodies Coelestial, Their Natures and Influences,* which, his subtitle claimed, was based on observations made "at leisure times" for more than thirty years. The book, which to modern readers has the appearance of total nonsense, was still being cited in the late 1800s.

In his youth, FitzRoy himself was a believer, but as he matured, he abandoned his belief in astrological weather signs and so-called lunarism. His reform came in part through discussions with Sir John Herschel, the man who named seven of the moons of Saturn and four of the moons of Uranus, well known for his own accomplishments but also as the son of the famous astronomer William Herschel. FitzRoy's reform came in part, too, from his reading of the work of others who strove to understand the wonders of the atmosphere. Over time, like many whose thinking is transformed with age, he became adamant about his views. Like a reformed smoker who has no patience with secondhand smoke, the reformed lunarist Robert FitzRoy had no patience with astrological meteorology.

But coincidences happen. Saxby occasionally got lucky. And with forecasts that were very general in nature—along the lines of "When the moon is full, there will be a storm, somewhere"—he could claim success on a regular basis. FitzRoy, driven by his circumstances and his personality, openly attacked Saxby's ideas. Saxby, in defending himself, publicly attacked FitzRoy. Their dispute became a matter of public interest. Newspapers as far away as Australia picked sides. Some readers supported FitzRoy and others supported Saxby, as though the two men were equals in the world of science.

The second man, the second critic, was Francis Galton, a half cousin and cheerleader of Charles Darwin. He was also a serious player in the scientific community—a well-connected polymath who, among other things, invented the field of eugenics and coined the term "nature and nurture" in the title of a book published in 1874. Galton had once proposed an invention for use by the Royal Navy called a "hand heliostat," a signaling device that relied on mirrors and a small telescope.

Galton's heliostat had been squarely rejected by an influential navy officer. That officer was, of course, a younger Robert FitzRoy. To make matters worse, when Galton requested data from FitzRoy—when Galton asked the official weather statist for official weather statistics—FitzRoy was less than cooperative. And so, justifiably or not, biased or not, Galton attacked FitzRoy's forecasts.

Within Galton's circle was the third critic, James Glaisher, a dashing character, a self-promoter, and an accomplished meteorologist who became known late in life for his exploration of the atmosphere via balloon. Glaisher did not object to forecasts themselves, but rather to free forecasts. As early as 1863, he and others were trying to commercialize weather forecasting, but they could not compete against an organization funded by taxpayers. So Glaisher, along with his followers, attacked FitzRoy's forecasts.

There were other critics, ready to cast disparaging remarks for the sheer fun of it or for their own advancement. There were political critics, people who saw FitzRoy's office as an extravagance, a waste of taxpayers' money. Even some shipowners—among those who might seem to have the most to gain from forecasts—were critical of FitzRoy. To at least some of them, forecasts of strong winds meant ships in port, ships waiting for a favorable forecast. Ships at sea earned money. Ships in port earned expenses. The owners critical of FitzRoy preferred the risks at sea to the risks in port.

FitzRoy, working hard to consolidate his knowledge, struggling to keep up with the statistics of the past and putting in extra effort to forecast the future, had to push himself even harder to defend his work. In 1863, while fighting to protect his ideas and his reputation, he published *The Weather Book: A Manual of Practical Meteorology.* "This popular work is intended for the many," FitzRoy wrote, "rather than for the few, with an earnest hope of its utility in daily life." He went on to complain about the state of meteorology, to lament the fact that the field had stalled, "had become statical in practice," nothing more than an exercise in the compilation of data.

FitzRoy sent a copy to John Herschel. "When you know that it was begun at Brighton," FitzRoy wrote of his book in an accompanying letter, "in my so-called holiday, on the 10th of last August, and was circulated to the public in December—your remark will probably be—'Better to have taken more time.'"

Herschel may well have thought just that. He discouraged FitzRoy from presenting his ideas to the Royal Society. If Fitz-Roy did so, Herschel thought, he would not be taken seriously. Despite his data, his finely tuned intuition, and his experience at sea, his forecasts were little better than guesses. The book was not grounded in theory. It was not underpinned by an understanding of the basic tenets of physics.

FitzRoy, throughout his life, suffered from dark periods. When he stopped working, when he paused his frantic pace, he grew despondent. He worried. Working as the nation's weather statist, the official weather watcher for Great Britain, he worried about criticism from the world of science, but also about personnel and budget issues. He worried about his personal finances, which were in disarray, and he worried about Darwin's views on natural selection, which conflicted with his own religious views. Others around him—family and friends, including Charles Darwin—commented on his melancholy.

Despite the popular success of *The Weather Book,* FitzRoy continued to face criticism. From a discussion about FitzRoy dated June 18, 1864, in the *Times,* the same paper that had published his first forecast just three years before: "What he professes, so far as we can divine the sense of his mysterious utterances, is to ascertain what is going on in the air some hundreds of miles from London by a diagram of the currents circulating in the metropolis." The *Times* repeated the words of Le Verrier's predecessor, François Arago: "Whatever may be the progress of the sciences, never will observers who are trustworthy, and careful of their reputations, venture to foretell the state of the weather."

As would be true of forecasts for the coming century, Fitz-Roy's forecasts were not very accurate. On average, they were no more accurate than they would have been had he simply said, "Tomorrow's weather will be similar to today's." He was fodder for those who sought to gain something from his failures or to entertain themselves with the shortcomings of his forecasts. The *Times* focused on the poor agreement between his forecasts and reality, while others focused on what they saw as the absence of a sound scientific foundation supporting his approach.

Besieged by critics, FitzRoy had to do something. Unlike Le Verrier, he did not dismantle his organization's ability to forecast weather. Instead, he killed himself. On April 30, 1865, disgraced, depressed, mocked by the *Times*, FitzRoy arose after a restless night and kissed his daughter as he passed into an adjoining room. Minutes later, he cut his own throat with a razor and bled to death in the bathroom of his home.

A monument on FitzRoy's grave, standing today in the shadow of apartment buildings and surrounded by a wrought iron fence, presents a quote from Ecclesiastes 1:6. "The wind goeth toward the south," it says, "and turneth about unto the north; it whirleth about continually, and the wind returneth again according to his circuits." FitzRoy lies interred beneath a stone suggesting that the wind shifts with a mind of its own, and yet he left this world not long before the emergence of a sound scientific foundation that would change forecasting forever.

## Chapter 3

# THEORISTS

Alone on deck, just after midnight on the seventh day of our voyage, I steer by the stars. I do not know the names of the stars by which I steer. I simply pick one that lies in the right direction and keep the boat pointed toward it, to the extent that the wind allows.

The right direction, at the moment, is south. We zigged and zagged to the east as far as we could, as far as was necessary, and two days ago we turned due south. Our first planned landfall, Isla Mujeres, lies 240 miles off.

The wind blows between five and ten knots, now from the east. I sniff for the scent of the sea but smell nothing that my mind can decipher. I listen to waves and wind, the sound of Beaufort's "gentle breeze."

Today is the last day covered by the forecast we received at the dock, but we may have picked up the northern edge of steady trade winds. If this Goldilocks wind stays with us, we are only two days away from Isla Mujeres.

At the helm, I listen to an audio recording of *Robinson Crusoe*. The young sailor and narrator, well before he begins his castaway island life, describes his first experience at sea. The wind rose and left him seasick and frightened. Conversing with a shipmate, he calls the wind a storm. "A storm, you fool you," says the shipmate. "Do you call that a storm? Why, it was nothing at all." It was, the shipmate says, nothing more than a capful of wind.

I eat a bar of chocolate. I watch the sky. I fight off sleep.

If FitzRoy's career as Meteorological Statist to the Board of Trade showed anything, if his experience as England's first forecaster could convey a message, it was this: To gain acceptance in Victorian England, to gain acceptance in Europe, forecasting would have to be grounded in scientific theory. It could not be based on anecdote or experience, not even the experience of the man who had once taken Charles Darwin around the globe.

The building blocks of the theory that FitzRoy needed were already in place, though he did not live to see them recognized. That honor went to others, and to understand it—to understand how scientific theory backed up by mathematics gained a foothold—requires a step back in time and eventually a trip across the Atlantic.

Among the early theorists of importance was the Greek thinker Archimedes, who realized that objects lighter than water—that is, less dense—would float, an insight that launched the field of hydrostatics, the science of fluids at rest. "Any object wholly or partially immersed in a fluid," Archimedes wrote sometime around 250 BC, "is buoyed up by a force equal to the weight of the fluid displaced by the object."

The atmosphere, of course, is a fluid, a sea of air. Any object, it turns out, could include a parcel of air within the atmosphere. Warm air—rarefied air, its molecules spread apart by virtue of temperature—floats above colder air. Moist air, carrying molecules of water that are lighter than the molecules of oxygen and nitrogen that make up most of the atmosphere, floats above dry air. Air rendered light by its warmth and moisture is buoyed upward by a force equal to the weight of the cold air it displaces.

Another theorist of importance was Leonhard Euler. Euler, in 1727, moved from hydrostatics to hydrodynamics by building on the ideas of men like Isaac Newton and Evangelista Torricelli. He moved, in other words, from understanding that a lighter body would float to understanding how fast the lighter body would float through a heavier body. With the work of Euler, it became possible, at least in principle, to understand how fast a body of lighter air—of air rendered less dense by virtue of its temperature or its moisture content—would move through a body of heavier air.

Although neither Archimedes nor Euler thought of himself as a meteorological theorist, each man's ideas when applied to the atmosphere were critically important. Not only did warm air rise, but it rose with a predictable speed. Understanding that reality was a crucial step toward understanding the mystery of wind.

Not all of the early theories were right. There was, for example, the theory advanced by Edmond Halley in the text that accompanied his trade winds map.

Like others who would come after him, Halley recognized the value of data networks. "It is not the work of one, or of few," he wrote in 1686, "but of a multitude of Observers, to bring together the experience requisite to compose a perfect and complete History of these Winds."

Using data gathered from ship captains, Halley described the trades, the winds that blew from Europe to the New World, the winds so important to commerce under sail. He mapped the specifics of where and when the winds blew. But he also wanted to know why they blew. "Wind," Halley wrote, "is most properly defined to be the Stream or Current of the Air, and where such Current is perpetual and fixt in its course, 'tis necessary that it proceed from a permanent unintermitting Cause." Even Halley, writing in 1686, well before FitzRoy's time, wanted to link the cause of the trade winds to underlying principles of physics, "to the known properties of the Elements of Air and Water, and the laws of the Motion of fluid Bodies."

Halley knew that warm air, "rarefied or expanded by heat," would rise "according to the Laws of Staticks." He knew that the sun would warm the air and set it in motion. In all of this, he was correct. But he went too far. He believed that the sun, rising in the east, chased the air around the planet, thereby driving the trade winds. "The presence of the sun," he wrote, "continually shifting to the Westwards, that part towards which the Air tends, by reason of the Rarefaction made by his greatest Meridian heat, is with him carried Westward and consequently the tendency of the whole Body of the lower Air is that way." His map was right, but his theory was dead wrong.

The ideas of two more theorists would correct Halley's mistake about the nature of the trade winds. The first was George Hadley, an English lawyer who dabbled in meteorology. In 1735, writing in the *Philosophical Transactions* of the Royal Society of London—the same journal that had carried Halley's trade wind maps and trade wind thoughts almost fifty years earlier—Hadley pointed out that air warmed by the sun and buoyed away from the surface of the earth would be replaced by cooler air from all directions. Despite Halley's assertion that the air would be driven forward by the warmth of the rising sun, flowing from east to west to fill the void left by air rendered buoyant by the new day's warmth, Hadley saw the obvious. "This rarefaction," he wrote, "will have no other Effect than to cause the Air to rush in from all Parts into the Part where 'tis most rarefied, especially from the North and South, where the Air is coolest, and not more from the East than the West, as is commonly supposed."

Something other than the rising and setting sun was needed to explain the trade winds. That something was the earth's rotation, coupled with the movement of air from the cooler regions to the warmer regions.

Warm air near the equator warms and becomes lighter. It moves upward. Cooler air moves in to fill the void. But the earth is a rotating sphere. The air on every part of the earth's surface is keeping pace with the earth's rotation, more or less. As Hadley put it, "Let us suppose the Air in every Part to keep an equal Pace with the Earth in its diurnal Motion." At the equator, the earth's circumference is about twenty-five thousand miles. A point on the equator covers twenty-five thousand miles each day, traveling at slightly more than one thousand miles per hour. A point on the north or south pole, on the axis of rotation, travels not at all. Strictly in terms of keeping up with the spinning earth, air at the poles spins in place, air at the midlatitudes moves fairly quickly, and air at the equator moves fastest of all.

Equatorial air, being heated, rises. Cooler air from the north and south moves across the earth's surface to fill the void. That cooler air, coming from higher latitudes and flowing toward the fast-moving equator, cannot keep up with the earth's spinning surface. As it approaches the equator, it is left behind. To an observer on the surface of the earth in the northern hemisphere, that cooler air blows from the east and northeast; it blows as the northeast trade wind.

It turns out that the earth's atmosphere, in very general terms, enjoys six distinct cells of rising and falling air, three in the northern hemisphere and three in the southern hemisphere. These cells, if the earth were entirely an ocean planet, a planet free of the interruptions caused by continents, would be far more regular and easily observed, but even on the earth as it is, they are real and regular enough to drive our weather. They are the Hadley cells, closest to the equator; the Ferrel cells, at the middle latitudes; and the polar cells, at the higher latitudes, close to the north and south poles. Each is accompanied by winds driven by the spinning of the earth, just as Hadley suggested.

But Hadley, though he was far closer to the mark than Halley had been, was not entirely right. In the language of physics, Hadley's ideas were based on the conservation of linear momentum, when he should have been conserving angular momentum.

Enter the second theorist, Gaspard-Gustave Coriolis, born in Paris in 1792. Coriolis was a physicist and eventually a professor of mechanics, but not, in his own mind, a meteorologist. In general terms, he was interested in energy transfer in rotating systems, and waterwheels were among the topics that drew his attention. Like others before him, he observed the apparent deflection of moving objects in a rotating system. He saw how objects moving from the inside of a spinning circle to the outside seemed to be left behind, how in traveling in a straight line they would appear, to anyone standing on the spinning circle and using that spinning circle as a frame of reference, to travel in a broad curve. Unlike others before him, he defined the apparent deflection mathematically. His equations conserved angular momentum.

Coriolis, coming well after Halley and Hadley, handed the world of meteorology a mathematical explanation of waterwheels that could be applied, by the right person at the right time, to the trade winds. But Coriolis himself was not the right person. He was not writing for meteorologists. He died in 1843, unrecognized in the world of meteorology until late in the nineteenth century.

Well to starboard, I see the lights of a ship. Ship lights come in green and red and white, and their number and arrangement say something about the ship and about its direction of travel. But red lights are always on the port side, and green lights are always on the starboard side.

I turn on our chart plotter, which stands just in front of the helm. The plotter's Automatic Identification System, its AIS, shows the ship on the screen. Through salty eyeglasses, I peer at the backlit screen, destroying my night vision. I punch a button, and the ship's identity comes up on the plotter, along with her speed and heading. She is a tanker heading east, toward Florida, at twelve knots. Her length is just shy of eight hundred feet. She carries what the AIS describes as "dangerous cargo."

The AIS warns me that we will come within one thousand feet of this vessel if we hold our course and she holds her course. We have less than ten minutes between now and our closest point of approach.

One thousand feet, onshore or in a channel in daylight, is one thousand feet. But to a fatigued and at best marginally competent captain more than a hundred miles from shore and sailing through darkness, one thousand feet is a near miss. For the tanker, it is less than two boat lengths. At twelve knots, the tanker travels one thousand feet in less than a minute.

I adjust course, falling off the wind, planning to pass behind the ship.

Suddenly I feel confused. I see a white light and a red light. On a sailboat, the white light is aft and the red light is forward, on the port side. The tanker's lights make it look as though my course correction will put *Rocinante* in front of her. I turn back to my original course.

I check the AIS. We are now on a collision course. In five minutes, if someone does not do something, *Rocinante* will never reach Mexico.

I awaken my co-captain. I start *Rocinante*'s engine. I am momentarily, stupidly, indecisive. We are in blue water, far from the shallow hazards of any coast. It would seem all but impossible for two boats to collide on a clear night in deep calm water, but we are in fact on a collision course. If I bear off, turning right, the ship's lights convince me that she will run us down. And if I head up, turning left, the AIS convinces me that the ship will likewise run us down.

Even with the engine running, a sailboat cannot ignore the wind. The wind in the sails limits maneuverability. It limits options. Even with the engine running, a sailboat is not a powerboat that can turn in any direction with impunity.

I head up, turning into the wind, but then follow through to bring *Rocinante* ignominiously about. The main and mizzen booms swing across the deck. My co-captain, abruptly awake and alert, works sheets and winches to bring the foresail across.

Now we sail away from the ship, assisted by *Rocinante*'s engine. On this course, northern Florida would be our next landfall. We travel entirely in the wrong direction, but we will not be run down.

Shaken, I look back, and I realize my mistake. Tankers carry white lights forward on masts, with red and green navigation lights aft. The AIS was right. The ship's lights were right. *Rocinante*'s captain was wrong.

My heart pounds.

I let the ship pass before bringing *Rocinante* about again. Back on course, we sail over the tanker's wake, the largest waves we have felt in at least three days.

Even as Coriolis described the apparent deflection of moving objects in a rotating system, an argument raged across the Atlantic, in the United States. Although no one knew it at the time, the argument was directly related to the ideas of Coriolis. It was an argument about what was known to some as "the theory of storms."

To understand the significance of that argument, ignore Coriolis for the moment, just as the world of meteorology ignored Coriolis. Instead, pay attention to James Espy, a recognized professional scientist, and William Redfield, a onetime saddle and harness maker who came to science relatively late in life.

James Espy, born in 1785 in central Pennsylvania, was a big name, known even to Le Verrier's old boss in France, who once compared Espy to Isaac Newton. Espy was employed for a time by the Franklin Institute, one of the nation's oldest science organizations, and like Benjamin Franklin he had ideas about weather. He observed smoke rising. He realized that warm air rises. He realized, too, that humidity buoys air upward. He recognized that warm, humid air cools as it rises. In cooling, the humidity goes from gas to liquid, from vapor to tiny droplets that join other tiny droplets that grow into larger droplets, as many as a million droplets joining to become a drop of rain that will eventually fall from the sky. This conversion from gas to liquid releases heat, a reality of physics first described in 1761. Espy realized that the heat released when gas becomes liquid, when water vapor turns into rain, further fuels the rising air. In the case of convection storms, he reasoned, water vapor turning into rain drives vertical winds.

Espy recognized, correctly, the driving force of thunderstorms. He saw how the storms grew, how they sucked in moisture, pulling it upward until the water vapor changed into water droplets, releasing heat, which further drove the upwardly mobile winds, which in turn sucked in more moist air from below. He saw how convection storms became more and more powerful, how they came from nothing to become awe inspiring and, often, terrifying.

He explained his ideas about convection storms to the American Philosophical Society in 1836, the year that FitzRoy and Darwin returned to England from their famous voyage. And in 1841, Espy described his ideas at length in a book, *The Philosophy of Storms*, which was 552 pages long, not including the eight-page preface or the forty-page introduction. Among many other things, the book includes a beautiful sketch of towering thunderclouds over a lake with mountains in the distant back-ground. In the sketch, arrows with their shafts slightly curved point upward and inward from below the clouds. Other arrows, near the top of the clouds, point upward and outward. At a glance, readers could see that air currents were moving verti-cally upward. With some concentration, with the dedication needed to follow the text, they could follow Espy's arguments about the inner workings of storms.

Espy knew that the rising air would leave a void, and he believed that the surrounding air would rush in from all direc-tions to fill the void. So far, he was correct, but he blundered on an important point. He believed that the air moving into the void would follow a straight path. Although Espy understood the inner workings of thunderstorms, he erred in envisioning sur-face winds as coming from all sides, sucked toward the center in straight lines, like wagon wheel spokes attached to a central axle.

By many accounts, Espy was argumentative. He belittled the work of others. He was, above all, self-confident, believing not only that he understood the workings of storms but that he could, through the judicious setting of forest fires, control storms. Unlike Benjamin Franklin, Espy was not the sort of man to modestly publish his findings in the upper left corner of another man's map. One of Franklin's grandchildren, who happened to know Espy, wrote about him: "His views were positive and his conclusions absolute. He was not prone to examine and reexam-ine premises and conclusions, but considered what had once been passed upon his judgment as finally settled." Another man wrote that Espy suffered from "want of prudence." And Presi-dent John Quincy Adams weighed in, calling Espy "methodi-cally monomaniac" and stating that "the dimensions of his organ of self-esteem have been swollen to the size of a goiter." In short, Espy's genius suffered from his arrogance.

Among those who did not succumb to Espy's bluster was William C. Redfield, the onetime saddle and harness maker and the son of a sailor lost at sea. Born in Connecticut, his varied résumé included time as a country merchant and a steamboat engineer. Never trained as a scientist, and busy pursuing a living, he nevertheless became an expert on two-hundred-million-year-old fish fossils. And he was the first president of the American Association for the Advancement of Science, which is now, more than a century and a half later, the world's largest science organization and the publisher of the prestigious journal *Science*.

In 1821, Redfield noticed something odd about the thousands of trees felled by a storm. Downed trees near Middletown, Connecticut, faced northwest, but seventy miles away the downed trees faced southeast. Ten years later, encouraged by a friend, he wrote about his trees. His observations, along with observations reported by others and compiled by Redfield, pointed to circular storms. Surface winds did not blow straight to a center point.

Redfield had the nerve to challenge Espy. "There appears but one satisfactory explanation of the phenomena," he wrote of his trees. "This storm was exhibited in the form of a great whirlwind."

Espy struck back, attacking Redfield and his ideas, saying he knew of no mechanism that would explain spiraling winds. Winds, Espy held, blew from high pressure toward low, following a straight line to fill the void left as hot air rose. Knowing of neither Hadley nor Coriolis, he saw no reason to believe that winds would blow in a circle.

While Espy drew supporters with his reputation and his passionate lectures, Redfield drew supporters with his observations. Not willing to ignore what they saw simply because Espy could not explain it, Redfield and his followers continued to report observations that refuted Espy's beliefs. Espy favored observations that fit his theories and ignored, or tried to ignore, those that did not. Redfield, in an 1839 letter published in response to Espy's theory, suggested that Espy's behavior was "not unlike that of him who in essaying to climb should commence at the last and highest step of the ladder."

The disagreements between Espy and Redfield became a matter of public spectacle, fodder for newspapers. The relationship was strained not just by the specifics of the theories but by strong beliefs in how science should progress, beliefs similar to those that would so dramatically impact FitzRoy's life. In Espy's world, hypotheses drove science. Espy wanted to explain why things happened the way they did. If he could not explain why storms spun, he could not believe that they spun at all. In Redfield's contrasting worldview, observations drove science. Redfield placed his faith in descriptive data, emphasizing what happened rather than why it happened. He saw that storms spun, and he was not especially troubled by his inability to explain why they spun. Tension between the two sides mounted.

In 1864, Joseph Henry, the head of the still young Smithsonian Institution, wrote of the dispute: "Two hypotheses as to the direction and progress of the wind in these storms have been advocated with an exhibition of feeling unusual in the discussion of a problem of purely scientific character." Henry compared the dispute about storms to storms themselves, "as if the violent commotions of the atmosphere induced a sympathetic effect in the minds of those who have attempted to study them."

But still, the scientific community sensed the possibility of reconciliation. "The interesting theories of Espy and of Redfield," said a report from as far away as Glasgow, "contradictory as they may now appear, will probably be found not incompatible with each other and they undoubtedly form the most important steps towards the widest generalizations which have yet been attempted in reference to the complex phenomena of the motions of the atmosphere."

In other words, both men might have something to offer. Maybe neither was exactly right and neither was completely wrong.

Reconciliation did come, but not from Espy or Redfield, nor from the Smithsonian, nor from anyone in the world of established science. Reconciliation came from a determined genius raised as a farm boy in rural Pennsylvania. It came when the farm boy, largely self-educated, took the dust-covered ideas of Archimedes and Euler and combined them with the ideas of Hadley and Coriolis. It brought with it the introduction of mathematics into the world of meteorology. And it brought with it a more refined view—a more correct view—of why the trade winds blow as they do.

That farm boy, possibly the most gifted meteorologist who ever lived, was William Ferrel.

William Ferrel was born poor and raised poor on a rural farm in south-central Pennsylvania, the eldest of eight children. After his death, his brother described the remains of their family home: "There is no town, not even a house to mark the spot that gave him birth; nothing but a mound of clay, the remains of the old stick chimney that belonged to the round log cabin that was covered with clapboard."

Growing up, Ferrel divided his time between his father's fields, his father's sawmill, and, for a short time, a one-room schoolhouse. Yet at the age of fifteen, having observed a solar eclipse, he calculated the timing of future eclipses. He once walked two days to purchase a book on geometry. He was twenty years old before he first read of the laws of gravity, and twenty-two before he had saved enough money to attend a school that offered instruction in algebra.

Out of school and living in western Missouri, he stumbled upon an abandoned copy of Newton's *Principia*. Later, via a traveling salesman, he ordered a copy of Pierre-Simon Laplace's five-volume *Celestial Mechanics*, the work that transformed our understanding of the solar system with mathematics. In both cases, Ferrel read through and mastered the books on his own. The printed word reached across the chasm of time and culture to join three master intellects—Newton, Laplace, and Ferrel.

When Ferrel was thirty years old, the Smithsonian Institution commissioned a report about storms. "In all our investigations respecting natural phenomena," the report proclaimed, "we assume that the operations of nature are subject to laws, and that these laws are uniform in their operation. A law of nature knows no exceptions. There is no place for science except upon this basis. Are storms subject to laws, and are these laws invariable?"

The words mirrored the mounting impatience with speculation in England and Europe, the sentiment that contributed to FitzRoy's suicide. Positivism—the idea that knowledge comes from combining observations of the real world with logic, preferably bolstered by mathematics—was gaining ground. Positivists wanted more than mere qualitative theory. They saw the need for absolute explanations, for quantitative theories, for governing laws, for statements along the lines of "if this, then that." They saw the need for mathematics.

Less than ten years after the Smithsonian report, in 1856, Ferrel published "An Essay on the Winds and the Currents of the Ocean" in, of all places, the *Nashville Journal of Medicine and Surgery*. It was an odd choice for an article about winds and currents, but Ferrel was a friend of the journal's editor. The editor had asked Ferrel to review a recent book about oceanography written by the well-known naval officer Matthew Fontaine Maury. Later, in a short autobiography, Ferrel wrote, "This I declined to do, but at length consented to furnish an essay on certain subjects treated in the book and notice Maury's views a little in an incidental way. This was the origin of my 'Essay on the Winds and the Currents of the Ocean.'"

The essay was reprinted as part of a series of meteorological papers in 1882.

If all the papers ever written on the world's weather were sorted in order of importance, Ferrel's might well rest near the top of the stack. Why? Because in this thirteen-page paper, Ferrel ended the arm-waving and speculation of his predecessors. Because while his contemporaries relied on vague and entirely incorrect allusions to the effects of electricity and magnetism, Ferrel explained the workings of the atmosphere in reasonable and mathematical terms. Here at last were the beginnings of the inviolable laws of nature that governed the atmosphere.

Dawn approaches from *Rocinante's* left, the sky turning from black to gray to cobalt to sky blue, and the stars, one by one, disappear. *Rocinante* and her crew are entirely alone, with no sign of other ships in any direction. The wind has all but died.

My co-captain sits next to me in the cockpit. She comments on the swells that have grown in the past hour, the low rolling waves from the east. *Rocinante's* booms swing when the swells roll under her hull. Her foresail slaps under its own weight. She makes no headway. The trade winds, if they were trade winds, abandoned us just before dawn.

I whistle softly. I try to persuade my co-captain to whistle.

Some sailors, past and present, rank whistling with sailing on a Friday, killing an albatross, carrying bananas aboard, and sailing with a man named Jonah. Others see whistling as a tool, a means of influencing nature. Sailors before me believed in their ability to whistle up a wind. They believed that the wind of their breath—resonating between their cheeks, exiting turbulently through pursed lips—could attract the wind needed to power their sails. I am with the latter group, the sailors who believe that whistling will summon a wind. I believe that if I whistle long enough, a wind is absolutely certain to come along.

In the old days, sailors lay becalmed at certain latitudes, starved of wind until they ran low on potable water. These men sailed before barometers and weather balloons and meteorological buoys and satellites. They sailed before William Ferrel. They did not understand why wind avoided certain latitudes, but they knew that it did.

They knew especially of the horse latitudes, at thirty degrees north and south of the equator. They did not know that this is the zone of sinking air that Hadley described, air that originates at the equator, air that rises with the warmth of the equatorial sun to a height of six or seven miles and then spreads to the north and the south, traveling a full thirty degrees of latitude before sinking back to the earth's surface.

Men traveling under sail, with limited supplies of water and food and patience, feared these horse latitudes as much as they feared storms. They feared them, but they sometimes had to cross them.

The horse latitudes, it is often but incorrectly said, were named for the need to kill off thirsty cargo. To save themselves, it is incorrectly said, becalmed sailors sometimes drove horses over the side, discarding them into the ocean rather than sharing the dwindling supply of potable water. More likely, the expression came from a traditional celebration that involved beating the effigy of a horse. In the days of long voyages under sail, sailors were often paid several weeks' or months' salary before leaving port. They called the first part of a voyage, when they were working for money that they had already received and probably already spent, "the dead horse time." Ships sailing west from England would often reach the horse latitudes just as the sailors' advances were worked off, and the sailors celebrated by parading a straw horse around the deck while beating it.

In the horse latitudes, men dependent on wind needed to feel as though they could influence their world, that they were not mere victims of an indifferent universe. Horse beaters prayed and cursed. Horse beaters whistled.

My co-captain, ever sensible, refuses to whistle. We drift, our sails limp, and I whistle alone.

In his thirteen-page paper, Ferrel translated Hadley's early discoveries about the movement of air into mathematics. He reminded readers that some parts of the atmosphere are lighter than others, less dense, either because they are warmer or because they hold moisture, that heavier air will flow under lighter air, forcing it upward, like water floating an inflated ball, and that the lighter air will spread outward at altitude. Air, moving north or south above a spinning earth, appears to deflect. "This is the same," Ferrel wrote, "as one of the forces contained in Laplace's general equations of the tides." He then gave an equation using the motion of the earth and the velocity of the moving air and the sines and cosines of latitudes. He called the equation "the analytical expression." He brought the ideas of Coriolis to bear on the atmosphere, the concept of angular momentum and the fact that air is subject to centrifugal force—an apparent force that seems to push rotating objects outward. Where Hadley's winds twisted to conserve linear momentum, Ferrel's twisted to conserve angular momentum. "If any part of the atmosphere has a relative eastern motion with regard to the earth's surface," he wrote, "this force is increased, and if it has a relative western motion, it is diminished."

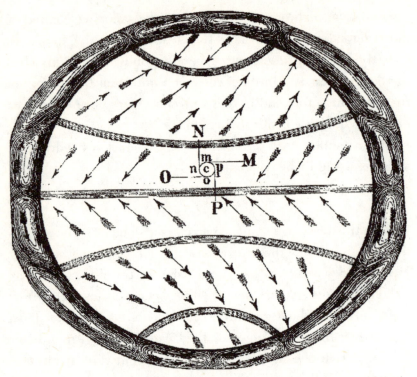

*The world's winds, as illustrated by William Ferrel in his 1856 paper, origi-nally published in the* Nashville Journal of Medicine and Surgery.

"Accordingly," he wrote, "we see that although Hadley's the-ory furnishes an explanation of the trade-winds, yet it does not account for many other remarkable phenomena in the motions of the atmosphere." Here he refers to differences in barometric pressure known from the increasingly abundant measurements coming in from all corners of the globe. He refers to phenomena like the high-pressure belt found close to thirty degrees, the home of the horse latitudes and the source of the trade winds, or, as he sometimes calls them, the "passage winds." The atmo-sphere, by virtue of the spinning earth, he surmised, would have a lumpy surface, with bulging lobes at certain latitudes. The atmosphere would not reach equilibrium, but would instead be concentrated at some latitudes and depressed at others. As air moved from high pressure toward low pressure, the earth would turn beneath it. If this moving air did not scrape across the ground or drag itself through trees and bushes or slosh across water surfaces, if there were no friction, it would, over time, end up on a trajectory exactly perpendicular to the pressure gradient. Air traveling long distances from high pressure toward low pressure, in the absence of friction, would never reach its low-pressure destination. This strange mathematical wind that would blow in the absence of friction became known as the geostrophic wind. Signs of it were everywhere, at scales ranging from the trade winds to the curving winds blowing around storms, but friction, a reality for all winds, changes the geostrophic wind's path, allowing it to take a curving route from high pressure toward low pressure, slowly moving toward equilibrium.

Every meteorologist should understand Ferrel's paper. Others need only understand its implications. Ferrel explained trade winds while also settling the argument between Espy and Redfield. Winds blew neither straight into the center of a storm nor in circles around the center, but in an inward spiral. The man born an impoverished farm boy—born with unusual genius, and by good fortune and fortitude exposed to the unusual genius of others—presented a picture of the earth's atmosphere that fit all of the facts known up to that time. What is more, he presented this picture mathematically.

*Chapter 4*

# INITIAL CONDITIONS

In a single step, in a paper originally published in a backwater medical journal, Ferrel set the stage for numerical forecasting, but he stopped short of actually proposing numerical forecasting. That honor would fall to yet another man, a Norwegian named Vilhelm Bjerknes. That honor would fall to a man who was prepared to combine the theoretical and mathematical world of those who thought like Ferrel with the practical world of those who thought like FitzRoy.

And key to the Norwegian's thinking, just as it had been key to FitzRoy's, was an understanding of initial conditions. Bjerknes knew he had to know something about the weather conditions today— the initial conditions—to forecast the weather conditions of tomorrow. Some of the tools used to measure those conditions were already available, while others would come later. But to understand what Bjerknes did, to understand the full impact of his contribution, one has to consider how we know about the initial conditions and to understand how overwhelming that knowledge has grown.

Still becalmed at midmorning on the eighth day of our crossing, *Rocinante* wallows in the swells, simultaneously riding the old swells from the east and new swells developing from the south. The eastern swells peak at two feet with a period of about four seconds, while the southern swells reach four feet, passing under us every six seconds. We move incessantly bow to stern and port to starboard, and in all directions in between. The motion conspires against appetite.

My whistling has not paid off. I have not whistled long enough. Though we are in constant motion, we make no progress. We make no headway.

On our satellite telephone, we receive a text message from a friend ashore. Text messages by satellite phone, like tweets, are the nearest living descendant of the telegram. Limited in length, they read like bulleted headlines. Properly worded, they imply more than they say. "Wind switching SW next 3 days," this one announces. "Consider diverting to Florida?"

We wallow 150 miles from Isla Mujeres, well to the north and slightly to the east of the island. Soon we will encounter the Yucatán Current, pushing northward, usually flowing at a knot or two or several. A wind from the southwest is an unusual wind in late autumn in this part of the Gulf of Mexico. And it is an inconvenient wind, a wind that will not take us where we had hoped to go. It is decidedly not a Goldilocks wind. But we sail — or do not sail — in the Gulf, and the Gulf's winds blow — or do not blow — with single-minded fickleness.

Before noon, the southwest wind arrives, light and sporadic at first, occasionally sending ripples up the south-facing slopes of swells, but growing to five knots and then ten before lunch. Our sails fill.

The wind blows directly from Isla Mujeres, as though originating there. We can tack toward Havana for a day or so, beating upwind, straining the crew and the boat, and then turn northeast for a while to beat in the other direction, zigzagging toward Isla Mujeres, sharing space with tankers and freighters in the shipping lanes just north of Cuba, making landfall in two or three days. Or we can divert toward Florida, aiming straight for the coast on a broad reach, the most comfortable and fastest tack of sailing, making landfall in two easy days.

We swing the helm to port, falling off the wind. We head to Florida.

The electric telegraph gave meteorologists a way to see the weather across large areas in something close to real time. Le Verrier in France and FitzRoy in England, seeing its potential for forecasting, were early adopters. In the United States, its potential was seen by those not so much interested in forecasting as in information that could advance science. By 1856—just twelve years after Morse's famous message to Vail asking "What hath God wrought?" and the same year that Ferrel published his paper in the *Nashville Journal of Medicine and Surgery*—the Smithsonian Institution had established a meteorological network and was posting a daily weather map on what amounted to a bulletin board. The map, displayed in the Smithsonian's Great Hall, used colored cards to indicate current conditions at various locations. Brown cards indicated clouds, black cards indicated rain, blue cards indicated snow, and white cards indicated fair skies. Arrows on pins could be turned to indicate wind direction.

Telegraphy, whether wrought by God or man, was a step toward the weather data communication network that has become one of humanity's most complex and beneficial achievements. Telegraphy allowed real-time weather data from multiple locations to be sent to a single central location where it could be compiled. But weather networks—telegraphic and otherwise—offer value only when they have information worth transmitting. Although temperature and humidity are important, too, from the perspective of wind the most useful information comes from two key instruments, the barometer and the anemometer, one to measure atmospheric pressure and the other to measure wind speed.

Belowdecks, mounted on the teak bulkhead separating *Roci-nante*'s galley from her forward cabin, a brass-encased barometer gives me air pressure. Wind is air moving from a region of higher pressure toward a region of lower pressure. When the barometer drops, air moves in. Wind blows. I can check the barometer every few hours to see if the pressure is rising or falling.

More accurately, if my barometer were in working order, it would give me air pressure measurements. If my barometer were in working order, I could check it every few hours to see if the pressure is rising or falling. In reality, my barometer neither rises nor falls. Its needle is stuck forever at 1,010 millibars. I tap it, and its needle vibrates, but it does not yield to changes in atmospheric pressure. *Rocinante*, like all great boats, is full of broken gear.

Back on deck, with my co-captain at the helm, I relax with the sense of speed. We may not be going where we intended to go, but we are going. We make five knots toward Florida.

The barometer, once known as the Torricellian tube, was invented in 1643. Its invention had nothing to do with weather. Its invention had to do with pumps used to move irrigation water and to clear flooded mines.

The Italian Evangelista Torricelli was born in 1608 and lived only thirty-nine years. During those thirty-nine years, he became a renowned mathematician, a maker of telescopes and microscopes, and the inventor of the barometer. In his thirties, Torricelli served for a short time as Galileo's amanuensis, tracking the great man's ideas just before he died. Among Galileo's many interests was the suction pump. Specifically, he was interested in why suction pumps could not lift water more than eighteen Florentine yards, or about thirty-four feet. He communicated with other scientists about this curious problem. Like children playing with soda straws, they realized that a tube filled with water and then blocked off at its top—a finger held over the top of the soda straw—would, when lifted vertically, retain the water. They built long soda straws. For example, sometime around 1640 a man named Gasparo Berti erected a three-story-tall straw made from lead, but with a glass globe fixed at its upper end. Berti's straw stood vertically above the streets of Rome, full of water, with its lower end open but immersed in an ornate vase. The water level in the straw, visible through the glass globe, dropped, but only to a point.

*Gasparo Berti's water-filled lead straw in Rome around 1640, with a glass globe attached to its top, demonstrated the weight of air. (Image from Wikimedia Commons)*

The sealed tube and the suction pump had something in common. Just as the suction pump functioned to a height of eighteen Florentine yards, the water trapped in the lead straw stood to a height of eighteen Florentine yards, but no higher. Beyond that height, a space formed at the top of the straw, plain to see in Berti's glass globe. The space looked empty—that is, it looked as though it held air. In fact, it held almost nothing. But for a few molecules that leaked upward from the water itself, the space held a vacuum.

Torricelli experimented with other liquids. One of these liquids was mercury. Mercury has its faults—it is poisonous, for example—but volume for volume it weighs more than thirteen times as much as water. Torricelli filled a long tube with mercury, closed the top of the tube, and immersed the bottom of the tube in a bowl of mercury. The level of the mercury in that tube dropped to something like thirty inches. Above that level sat emptiness, a vacuum similar to the vacuum that formed above a much taller tube of water. The empty space at the top of the tube became known as a Torricellian vacuum, and the tube—a barometer—became known as a Torricellian tube.

Galileo, before he died, knew that air has weight, establishing once and for all that air is a substance in and of itself—that air, despite being transparent, is something rather than nothing. Torricelli, before he died, showed that the weight of the air pressing down on the liquid pool at the bottom of his tube— that is, the weight of the atmosphere—is about the same as the weight of thirty inches of mercury or eighteen Florentine yards of water. It is the weight of the atmosphere that elevates the mercury or water in closed-ended straws.

Torricelli noticed something else. He noticed that the level of mercury in his Torricellian tube changed over time. The weight of the atmosphere is not constant. In 1644, he wrote a letter to another Italian, Michelangelo Ricci, describing "an instrument

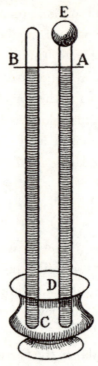

*Torricelli's barometer, sketched in a 1644 letter to his friend the Italian mathematician Cardinal Michelangelo Ricci. Torricelli wrote, "We live submerged at the bottom of an ocean of the element air, which by unquestioned experiments is known to have weight."*

that would show the changes of air, now heavier and denser, now lighter and thinner." His mercury-filled tube was that instrument.

The link between wind and the height of mercury was not immediately apparent, but there were observations from scattered sources, comments from scientists feeling their way into a new area of knowledge. Scientist Robert Hooke wrote a letter to scientist Robert Boyle, the man who discovered the relationship between pressure and volume in the atmosphere, on October 6, 1664: "I have also constantly observed the baroscopical index (the contrivance, I suppose, you may remember, which shows the small variations of the air) and have found it most certainly to predict rainy and cloudy weather." When the mercury sank, rain was on the way, and when it rose, clear weather would follow. But two weeks later he wrote again, describing inconsistencies in his observations. The height of the mercury changed "notwithstanding the variety of winds."

The mercury barometer faced another problem. It did not work at sea. Subject to the never-ending motion of a ship, the mercury sloshed around in the bowl at the base of the tube and moved up and down in the tube itself. Modifications were made to dampen the motion. In particular, the middle of the tube was constricted, limiting flow, but mercury barometers still did not work well afloat.

Yet another scientist, the Frenchman Lucien Vidi, developed the aneroid barometer in 1844. The aneroid barometer does not rely on mercury. Instead, as the pressure changes, a sealed capsule changes shape. The changing shape of the sealed capsule is transmitted to a needle that moves across a scale. Aneroid barometers became commonplace and remain commonplace. Even in today's world of electronic instrumentation, aneroid barometers, frequently in decorative brass housings, can be seen on desks and mantels.

The aneroid barometer, along with the mercury barometer, supplied information for the telegraphic networks springing up in the United States, Europe, and England. This information, read from a single location, could be confusing. The correlation with current and future winds could be inconsistent. But with information from dozens or even hundreds of locations, and with this information laid out on maps, a very compelling story began to emerge. When air pressure and wind data were overlaid on a map, only the least observant could fail to see a relationship between the two. Hooke may have seen inconsistencies at a single location, but as a general matter, a falling barometer foretells future winds. Air moves from places where its molecules are tightly packed toward places where they are widely spaced, something like a crowd surging out of a stadium, with its course deflected over long distances by virtue of a spinning earth.

I sit in the sun on deck, afloat at the bottom of a sea. Thanks to Galileo, Torricelli, and their successors, I know that this sea of air in its entirety weighs more than five quadrillion tons. Five quadrillion is a five followed by fifteen zeros. More than three-quarters of that hefty weight lies below about eight miles of altitude, in the zone of the atmosphere that makes weather. The weight, of course, feels weightless because it pushes equally in all directions. But when the wind blows, I feel an expression of the weight of air. When the wind blows, the atmosphere no longer pushes equally in all directions. Its weight is greatest from windward.

There in the sun, I read FitzRoy's *Barometer and Weather Guide,* the third edition, written for his employer, the Board of Trade, in 1859, the same year he launched his telegraphic network. "I think that the neglect of the use of the barometer has led to the loss of many ships," FitzRoy wrote. "From a want of attention to the barometer, they have either closed the land (if at sea), or have put to sea (being in harbour in safety) at improper times; and in consequence of such want of precaution the ships have been lost, owing to bad weather coming on suddenly, which might have been avoided had proper attention been paid to that very simple instrument."

The main point of FitzRoy's *Barometer and Weather Guide* was to advocate proper use of the barometer. Not everyone in his time understood the value of the barometer, or its limitations. From the third edition of FitzRoy's volume on the barometer, released just before the wreck of the *Royal Charter:* "No violent wind will blow without the barometer giving warning." Had the *Royal Charter*'s crew watched its barometer, they might never have left Ireland. They might have survived.

"Aneroid barometers," FitzRoy wrote, "if often compared with good mercurial columns, are similar in their indications, and valuable; but it must be remembered that they are not independent instruments; that they are set originally by a barometer, require adjustment occasionally, and may deteriorate in time, though slowly."

*Rocinante*'s barometer, frozen forever at 1,010 millibars, is an aneroid barometer. More accurately, it is a badly corroded aneroid barometer.

FitzRoy did not rely on the barometer alone. He advocated combining information from the barometer with the appearance of the sky—the shape and texture of clouds, for example. With his background, with his time at sea, he likely noticed signs instinctively. He may have sensed signs that he could not articulate.

When I finish his small book, I check the wind speed. I take comfort in the fact that *Rocinante*'s anemometer—her wind meter, called an anemometer after the Greek word for wind—works for now. Atop *Rocinante*'s mast, aloft, above all of her sails and but for a short antenna the highest point on the boat, the three cups of the anemometer spin. Here in the cockpit, a digital readout displays eight knots.

But *Rocinante*'s anemometer works only to a point. The wind speed is not in fact eight knots. We travel at five knots, with the wind not quite behind us, but behind and to the starboard side. We sail on a broad reach, with the wind coming from an angle more or less behind the boat. Without overcomplicating matters, we have to add most of the boat's speed to the reading from the anemometer. The apparent wind, the only wind we can measure from the moving boat, blows at eight knots, but the true wind speed blows at just under twelve knots. We sail toward Florida in what Rear Admiral Beaufort would identify as a moderate breeze.

I watch a school of flying fish dash away from *Rocinante*, appearing from the sides and crests of waves, shimmering silver in the sun, pectoral fins extended as wings, as lovely as a flock of birds. They launch from the tops of waves. They can accelerate in midair by dipping a frantically moving tail fin into the water. In a recorded flight off the coast of Japan, a flying fish once stayed aloft for forty-five seconds. Speeds of close to forty knots have been clocked, and glide distances of a quarter of a mile have been measured. The fish take advantage of updrafts coming from the water's surface, of turbulence that rolls off the backs of their wing-like pectorals, only to be trapped between them and the water, forming an air cushion of sorts, a pillow. Pilots experience that same pillow of air and call it ground effect or float—float, even on jets as big as the Airbus A380-800 or the Boeing 747, because pilots nearing the ground feel resistance to further descent, as if their planes were reluctant to give up their airborne status.

I put FitzRoy down and go below to tap again on the face of *Rocinante's* broken barometer. The needle vibrates with each tap but reads a steady 1,010 millibars.

The challenge of measuring wind speed comes from the simple fact that wind cannot be tagged. Moving air is as transparent as still air. One cannot watch the wind move.

When Rear Admiral Beaufort and his predecessors came up with their wind scales, there were no convenient options for consistently measuring wind speed. Observers feeding the earliest data networks—the networks of Henry and Le Verrier and FitzRoy—generally were not equipped with the simple spinning wind meter of today, the cup anemometer. An observer, whether at sea or on land, did the best he or she could. When an observer in Mansfield gave a different wind speed than an observer in Birmingham or an observer in Bristol, no one knew whether the wind was blowing at different speeds or the observers were simply out of calibration. When men like Espy and Redfield argued about the nature of the winds that blew in and around storm centers, there was a fair amount of guesswork and bluster when it came to describing the speed at which the winds blew.

The first anemometers, no more sophisticated than a light ball hanging from a string, were around in the fifteenth century. A breeze blowing against the ball moved it away from the resting position, holding the string at an angle. That angle offered observers a rough estimate of wind speed. If the wind blew fast enough, the ball might be held parallel to the ground. Stronger still, and the ball might be blown away entirely, lost, no longer measuring anything at all.

That simple device was invented more than once. A more sophisticated version used a disk hanging from an arm that turned on an axis, acting something like a needle on a gauge. Robert Hooke is often credited with the original version, but in fact it had been around for about two centuries by the time he was born. Leonardo da Vinci, dreaming of human flight and considering how aviators might one day take advantage of breezes, sketched a simple anemometer sometime around 1480. Next to his sketch he wrote, "For measuring distance traversed per hour with the force of the wind." The sketch was based on a thirty-year-old design put forth by the Italian architect Leon Battista Alberti, who may or may not have been the first to come up with what was, after all, a very simple idea.

The devices of Alberti, da Vinci, and Hooke suffered from problems with precision. Two readings of the same wind did not yield the same value. If the device was pointed directly into the wind, its hanging disk moved farther than it would if the device was angled to the wind. A wind believed to blow at ten miles per hour might in fact have been blowing at twelve miles per hour or fifteen miles per hour. Alberti's device, the swinging disk anemometer, was clumsy and inelegant.

The cup anemometer, the modern anemometer sitting on my masthead, was invented in 1846 by John Thomas Romney

Robinson, a Dubliner once recognized for cataloging 5,345 stars. His whirling cups caught on quickly. Robinson claimed that the speed at which the cups whirled was independent of the size and shape of the cups and the arms that attached them to their spinning axis. He believed that the rate at which the cups spun could be simply converted to a reliable wind speed. Forty-three years later, in 1889, after having his claims verified by no less an organization than the Royal Meteorological Society, he was shown to be wrong. With Robinson long dead, the refutation was published in the journal *Science,* in a one-page letter to the editor: "It would appear that the different conclusions reached by the wind force committee of the Royal Meteorological Society in their open-air experiments are largely misleading and in error." The point of the letter was not that the cup anemometer should be abandoned, but that it required calibration. It had to be fine-tuned in known wind speeds to give accurate measurements.

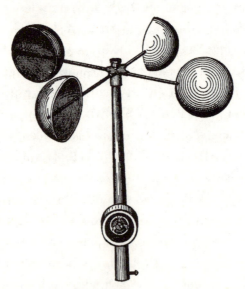

*Robinson's cup anemometer, invented in 1846, did not give perfectly consistent wind speed measurements, but it was a dramatic improvement over earlier anemometers. Drawing reproduced from Frank Waldo's 1896 book* Elementary Meteorology for High Schools and Colleges.

Robinson's anemometer, although commonly used today, did not lead to the abandonment of other approaches to measuring wind speed. More than a century after Robinson, almost six centuries after Alberti, improvements continue and new devices appear. The "electrogasdynamic spectral anemometer" was patented in 1973, the "ultrasonic anemometer" in 1979, the "anemometer having a graphite fiber hot wire" in 1987, and the "micro gust thermal anemometer" in 2006. Measuring the speed of the wind, it turns out, is not quite as simple as the spinning cups on top of a masthead would suggest.

Late in the afternoon the wind dies, petering out to nothing, and the anemometer on the masthead rests. *Rocinante* herself is at rest. The sea is calm. Six pantropical spotted dolphins approach from the north, leaping and splashing around our slumbering boat. With the aid of a bucket and a short piece of rope, we bathe on deck and then take turns swimming in the open ocean. Afterward, on board, I admire my co-captain as she lets her hair and skin soak in the warm air while she inhales the sunshine.

According to our charts, we have ten thousand feet of water under the keel. A few flat blades of turtle grass drift alongside, dark green strips of vegetation against the dark blue of deep ocean, visitors torn away from the shallows of Cuba or Mexico or Florida.

Evening approaches and the wind reappears, now from the southeast, blowing up the gutter between Cuba and southern Florida. The anemometer spins. *Rocinante* moves through the water.

The wind builds. The anemometer shows fifteen knots, and then twenty, and then a gust of twenty-five. We reef our mainsail, reducing its size by one-third.

Four-foot waves, steep-sided and sharp-crested, build quickly. Now and then, water splashes into the cockpit. One wave sends a gallon of the Gulf of Mexico through the open companionway and onto the teak-and-holly deck of *Rocinante's* salon. My co-captain takes the helm while I mop, and we close the companionway hatch as the sun sets.

I read twenty-nine knots on the anemometer, the low end of Beaufort's near gale. Taken at face value, the average speed of the air molecules speeding past, moving from high pressure toward low pressure, twisted in their path by a spinning earth and tripped by friction, is twenty-nine knots, but the wind does not blow steadily at that speed. In gusts, cup anemometers are prone to something meteorologists call overspeeding: they spin too fast in turbulent winds. More important, on this tack *Rocinante's* speed adds a knot or two to the speed of the wind. With that and a short-lived gust, maybe combined with the roll of the boat, the anemometer claims to experience a near gale when in fact it experiences no more than a strong breeze or even a fresh breeze.

Like sailors prone to exaggeration, my anemometer reports a stronger wind than nature provides. The apparent wind, aboard a moving boat, is seldom the true wind. Nevertheless, it is the apparent wind that the boat experiences. It is the apparent wind that is felt by both of *Rocinante's* captains. I reef the foresail before darkness blankets the sea.

Soon after, a ship's lights appear well to starboard. I see only its white light and a green running light. On my electronic plotter, I see that it is a freighter, bound for Cartagena, Colombia. In the night, we pass starboard to starboard, three miles of ocean between us.

Halley, in drawing his trade winds map, relied heavily on reports from ships. He was neither the first to do so nor the last. Forecasters, hungry to understand initial conditions at any one time, need data. On sea as on land, they need networks of observers.

Until 1936, forecasters depended entirely on observations provided by various navies and by volunteers on merchant ships.

Then Britain's Meteorological Office, FitzRoy's legacy, installed a professional meteorologist on a cargo steamer plying the North Atlantic trade route. Two years later, the French converted a merchant vessel into the first weather ship, assigned to find a location in the North Atlantic, stay there, and observe the weather.

By World War II, the advantages of stationary weather ships were apparent. Navy vessels were assigned to maintain positions at sea. They provided observations, but they also made good targets.

After the war, the Ocean Weather Ship agreement called for the United States, Canada, France, the United Kingdom, Norway, Sweden, Holland, and Belgium to provide weather ships and share data. The converted British corvette *Marguerite,* two hundred feet long, steam-powered, and ingloriously renamed *Weather Observer,* sailed into the North Atlantic on August 1, 1947, becoming the program's first ship. She sailed just in time to catch the winter storms. In less than a year, she was joined by the *Weather Recorder,* the *Weather Watcher,* and the *Weather Explorer.*

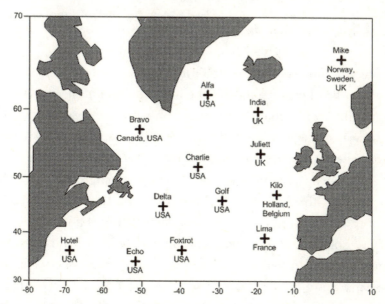

*Positions of weather ships agreed on at the International Civil Aviation Organization's 1946 conference in London. (Image from Wikimedia Commons)*

Weather ship sea duty involved steaming back and forth within a half mile of an assigned location for twenty-one days at a time. Unlike normal ships, a weather ship on duty had nowhere to go. Unlike normal ships, a weather ship on duty took no action whatsoever to avoid bad weather.

*The weather ship* Weather Adviser, *later rechristened the* Admiral FitzRoy. *(Published with the permission of Derek Ogle and the assistance of Ocean Weather Ships and Paul Brooker)*

Storms punctuated the tedium of going nowhere at sea. "I saw the Northern Lights and the biggest waves I have ever seen in my life," remembered one man of his time on a weather ship. "The waves came over the bridge. I still remember the way the ship rolled when we stopped engines." When the weather was fine, he filled the long hours on watch by pacing back and forth on the ship's bridge. "I walked the bridge on the 12–4 shift," he later recalled. "It took me 55 seconds or 61 times an hour."

The ships offered services beyond mere weather recording. They assisted in search and rescue. They maintained homing beacons to guide aircraft. They towed nets to collect plankton samples. Acting on a request from Cambridge University, one of them released twenty shearwaters—seabirds with long wings—

far out at sea. Eighteen of the birds returned to their nests ashore. The first to arrive traveled 450 nautical miles in thirty-six hours.

Budget shortfalls sank the US weather ships by 1974. Other nations followed their lead. In 2009, a headline in the science magazine *Nature News* read, "Last Weather Ship Faces Closure." The *Polarfront,* funded by the Norwegians, would no longer hold its position at what was known as Station Mike, two hundred miles off the coast of Norway at the latitude of northern Iceland.

Researchers complained. "It's a blow," said one. Another worried about "dramatic negative consequences."

The director of the Norwegian Meteorological Institute responded. "I have received an overwhelming number of statements stressing the importance of continuing the operation of the ship," he said, "but no further offers to share the expenses." The *Polarfront* steamed away from Station Mike on January 1, 2010.

Even before the demise of the weather ships, reports from other vessels that were going about the business of fishing or commerce, commonly called vessels of opportunity, became important, and they remain important today. The National Oceanic and Atmospheric Administration, the United States' weather watcher, offers these words to mariners: "Only YOU know the weather at your position." The government's request for data from vessels of opportunity, to be reported at six-hour intervals, follows that simple statement.

The national effort is part of the much larger World Meteorological Organization's Voluntary Observing Ships program. In the 1980s, 7,700 ships reported in to the program. By 1994, the number had dropped to 7,000. Today, only about 4,000 ships report. Even though the meteorological gear is installed aboard at no cost to the ship, and even though transmissions are sent at no cost to the ship, and even though the world's oceans are full of areas where data are sparse, other priorities have displaced voluntary weather reporting. Crew sizes have shrunk, reducing spare time that could be devoted to voluntary reporting. Among those ships reporting, not all report at all of their designated times. Among those reporting at their designated times, not all reports are accurate. The Voluntary Observing Ships program has fallen on hard times.

While the weather ships fell from favor and went extinct, while the Voluntary Observing Ships program grew and then shrank, another program emerged. By 1951, the United States was deploying automated weather buoys. Automated weather buoys cannot do everything that was done by ships. They cannot, for example, release shearwaters. Nor can they take repeated water samples at different depths or launch weather balloons. Nor can they assist with search and rescue. But they are inexpensive compared with ships.

The early automated buoys, called NOMADs, for Navy Oceanographic Meteorological Automatic Devices, were twenty feet long, with aluminum hulls. NOMADs, despite their name, were not nomadic. They were anchored. They were designed to withstand storms. The designers take pride in their claim that no NOMAD has, to date, capsized.

NOMADs have been joined by other moored, or anchored, buoys. There are the three-, ten-, and twelve-meter discus buoys and the smaller coastal buoys, all circular platforms sporting masts with instruments. In shallow water, they are held in place by chains. In deeper water, the anchor line, which can approach two miles in length, can be a combination of chain, nylon, and buoyant polypropylene.

And there are drifters, unanchored buoys made from fiberglass or plastic, measuring less than two feet in diameter, each towing a small drogue—an underwater parachute—to slow its passage through the water. A few hundred new drifters go over the side each year, launched by research ships, freighters, tankers, naval aircraft, and, occasionally, sailboats. At any one time, more than a thousand of them are adrift. They can be found in all of the world's oceans, never complaining about heat or cold or tedium or wave height, never suffering from seasickness, never sending out distress signals. When they fail, they fail quietly.

Moored buoys and drifters send data skyward, to satellites. Moored buoys and drifters, along with reports from ships, oil platforms, and almost any willing source afloat, offer a synoptic view of weather at sea that could not have been imagined by the likes of Halley, FitzRoy, Le Verrier, and Richardson. But still, it is not enough. If forecasters had no other information on conditions at sea, if they had no data other than that coming in from ships and buoys, their forecasts would suffer along with their reputations and, perhaps, their self-esteem.

As night progresses, the wind lays down, stabilizing somewhere near ten knots, Beaufort's gentle breeze. *Rocinante* all but sails herself. I recline in the cockpit, one toe on the helm, utterly relaxed but fully awake. Clear but moonless, the sky glows with the Milky Way and countless points of light. Shooting stars descend every few minutes, sudden streaks of light, exciting but short-lived. I listen to water passing under *Rocinante's* hull and to waves lapping alongside.

In the darkness on board, I see shadowy forms in the cockpit — a winch, the helm itself, the companionway. But the veil of night hides everything beyond the boat's rails. Our sails and rigging reside in darkness. I stare at the sky, and after a time I am rewarded by a streak of light so bright that it illuminates sails and deck, momentarily chasing away the shadowy world, burning not white but orange and yellow and red as it dashes in a trajectory over Cuba, to starboard, not just a shooting star but a fireball. It burns for what may be as long as two seconds, temporarily destroying my night vision, and then it fades.

Later, my night vision restored, the cockpit reduced again to shadowy forms, I watch a satellite, a dim point of light following a straight line across the sky, moving against the static background of stars. *Rocinante* continues to all but sail herself, my role limited to a single toe on the helm, offering, now and then, the slightest adjustment to her rudder as she moves across the sea.

Before there were weather satellites, balloons and aircraft gave meteorologists something of an aerial view. For example, Fitz-Roy's critic James Glaisher, one of the founders of the Royal Meteorological Society in 1850, rode balloons into the sky in pursuit of the society's lofty mission, which was nothing less than understanding the laws of climate and weather. On his most memorable flight, somewhere close to thirty thousand feet, a caged pigeon along for the ride lost consciousness. Glaisher, too, lost consciousness. His still conscious assistant struggled to release gas from the balloon. The assistant's hands at that point in the flight were stiffened by cold and no longer functional, so he pulled a dump cord with his teeth. The balloon lost altitude. Glaisher recovered. The caged pigeon was less fortunate.

Glaisher introduced others to ballooning. On one ascent, Glaisher took twenty-eight passengers up in a balloon called the *Captive*. The *Captive* was held to the ground by two thousand feet of anchor cable. The wind speed, according to one account, was sixty miles per hour, Beaufort's whole gale strength in his 1831 scale, one notch up from a strong gale and one notch down from a storm. Even allowing for exaggeration, the wind speed exceeded levels acceptable to the prudent balloonist.

One of the passengers wrote about the flight: "The strong wind whistled through the ropes, the balloon lay over, and the car oscillated violently." According to other accounts from the *Captive*, Glaisher paid no attention to the howling wind or the swinging car with its twenty-eight other passengers. His eyes, according to one passenger's account, remained locked on his instruments. It is a matter of opinion whether Glaisher was courageous, foolish, or suicidal.

Glaisher knew, from his ascents, that the wind speed on the ground might not indicate the wind speed aloft. The wind direction, too, might differ. "The balloon in almost every ascent was under the influence of currents of air in different directions," he wrote. "Sometimes directly opposite currents were met with at different heights in the same ascent, and three or four streams of air more than once were encountered moving in different directions." From his balloon, Glaisher saw what Benjamin Franklin had seen in the movement of storms.

By 1896, when Glaisher was well into his eighties, unmanned weather balloons had contributed to the discovery of the tropopause and the stratosphere, layers of the atmosphere that occur five to ten miles above the earth's surface, beyond the reach of what is normally thought of as weather. In the tropopause, moisture disappears and temperature stabilizes. Increasing altitude no longer decreases temperature. Updrafts and downdrafts disappear. The tropopause and the lower stratosphere are sweet spots for commercial jet traffic, areas where seat belt lights might be turned off and passengers might be welcomed to wander about the cabin.

Today, meteorologists launch something like sixteen hundred weather balloons each day. Unlike the early balloons, which provided useful information only after they sank back down to earth to be recovered by their owners, modern balloons stream data earthward.

More information streams earthward from commercial airliners. Alaska Airlines, Delta, United, Air France, Lufthansa, Jetstar, Qantas—virtually all the well-known carriers—share data through a system similar to the Voluntary Observing Ships program. It is called AMDAR, for Aircraft Meteorological Data Relay. Like its sister program for ships, it is run by the World Meteorological Organization.

The seventy-nine-page AMDAR manual starts with three lines of boxed text, quoting James Boswell's 1791 biography *The Life of Samuel Johnson*: "Knowledge is of two kinds. We know a subject ourselves, or we know where we can find information upon it." AMDAR, harvesting information from airplanes, kicks out something like three hundred thousand observations each day.

Still more information streams earthward from satellites. The first weather satellite, Vanguard 2, a pretty shining sphere a mere twenty inches in diameter, flew into space on February 17, 1959. It outdid many of its predecessors in that it did not die in flames on the launchpad or tumble out of the sky without achieving orbit.

Vanguard 2 carried, among other things, two optical scanners. Each was not much more than a tiny telescope focused on a solar cell. When light hit the solar cell, electrical current flowed. As little Vanguard 2 circled the earth, it looked at clouds and ground and ocean. Clouds reflected more light and generated more current than the ground, and the ground reflected more light and generated more current than the sea. The satellite recorded data to a tape, and now and again the tape's contents were transmitted home, after which the tape was rewound and reused. The satellite, in other words, told operators about cloud cover.

All was well but for two things. First, the data were numerical, not visual, not the sort of thing that would play well on the black-and-white televisions perched in American living rooms. Second, the satellite's orientation was off. The optical scanners did not quite point where they were supposed to.

Vanguard 2 remains in orbit, no longer transmitting, now a derelict satellite, just one more piece of space junk. If all goes well—if Vanguard 2 does not, for example, collide with another piece of space junk—the satellite will continue to orbit for three centuries, silent and shining, a ghost ship in the sky.

On April Fools' Day 1960, thirteen months after the Vanguard 2 launch, TIROS-1 found its way into orbit. TIROS-1 was a forty-two-inch-wide, twenty-two-inch-tall can weighing, on earth, 283 pounds. Newspaper accounts referred to it as a "moonlet." It carried more than solar cells. As implied by its name—TIROS, for Television Infrared Observation Satellite—it also carried television cameras. During the seventy-seven days that it functioned, it sent back 19,389 black-and-white images of clouds. The movement of air, the wind, could be seen in the shapes of those clouds and in their motion over time. The patterns argued over by men like Espy and Redfield showed up in black and white.

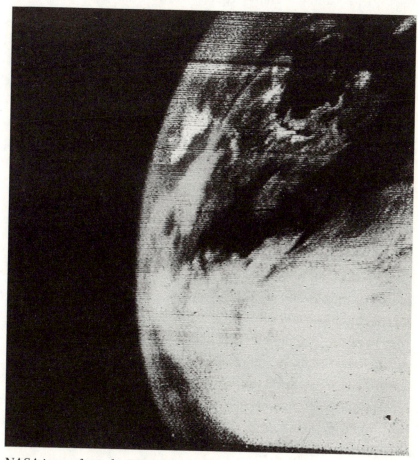

*NASA image from the TIROS-1 satellite, April 1, 1960.*

Not long after the launch, Harry Wexler, the chief meteorologist of the Weather Bureau, said that TIROS "established the feasibility of space weather stations — no question about it."

Over the next few years, nine more TIROS satellites found their way into orbit. By 1966, an approach that relied on lone weather satellites had been replaced by one using weather satellite systems that provided ongoing input to forecasters. There was Nimbus and ESSA and ITOS and GOES. Before long, weather satellites evolved into environmental satellites, earth-observing systems fired into space to look backward, to inform those below about exactly what was going on. Now there are polar satellites, circling north and south. There are geosynchronous satellites, traveling at a speed matching the earth's rotation and returning, from the viewpoint of an earthbound observer, to the same point in the sky every night. And there are geostationary satellites — a special class of geosynchronous satellites perpetually hovering over a fixed point on the equator, sailing in windless space 22,236 miles above the earth's surface and making good headway at just under six thousand knots. Weather systems are observed in action every day, all day.

By 1987, satellite technologies made it possible to measure surface wind speed over the ocean from space. Today, some would argue that surface measurements of wind, the measurements collected by ships and sea buoys, serve mainly to calibrate the measurements taken from space.

The satellites determine wind speed in much the same way that FitzRoy once determined wind speed. When FitzRoy commanded the *Beagle,* carrying Darwin across the globe, he recorded sea state as a proxy for the wind, using Rear Admiral Beaufort's scale, since he could not reliably measure wind speed itself from his moving vessel. Satellites today also measure wind speed on the world's oceans based on the state of the seas' surface, using a method tested on the Skylab space station in 1973. Scatterometers on satellites send out pulses of microwaves. The microwaves reflect differently off rough seas than off calm ones. The satellites, sensing the reflected microwaves, use this information to compute wind speed. FitzRoy and Beaufort, if they were alive today, would smile at the news.

All of which brings the story back to the Norwegian Vilhelm Bjerknes. In 1904, Bjerknes suggested a means of combining the knowledge of today's weather with theories about the workings of the atmosphere that would be further developed by Richardson, leading to the world's first numerical forecast — that is, the world's first modern forecast. Bjerknes's suggestion pointed to a way of using all the information that, even then, was streaming in on telegraph wires. His suggestion still made sense a hundred years later, when data flowed from ships and buoys and drifters and airplanes and satellites, a never-ending stream of information that staggers even the largest of imaginations, information that is not only useful but critically important to anyone hoping to understand tomorrow's weather today, but that at the same time turns out, for the purposes of long-range forecasting, to be entirely insufficient.

*Chapter 5*

# THE NUMBERS

Before dawn, with ten knots of apparent wind, with *Rocinante* traveling at five knots toward Florida, we make soundings. That is, for the first time in more than a week, our depth sounder finds the bottom.

Not long ago, before the invention of sonar, soundings were found with a lead line. A sailor would cast a lead weight attached to a thin rope over the side of the ship. If the sailor felt the weight hit the bottom, the ship had made soundings. Marks on the line indicated the depth. A ship that made soundings was a ship that would watch for a shoreline.

The word itself, in the context of making soundings, has nothing to do with sound. It probably comes from the Old English word "sund," meaning something like "water" or "sea," as in Long Island Sound or Puget Sound. The fact that we use sound waves to probe the depths today is nothing but coincidence. The fact that the word "sound," as in "sounding the depths," sounds like the word "sound," as in "the sound made by the depth sounder," is a chance occurrence.

Lewis Fry Richardson was one of the early contributors to sonar systems. Although a decade would pass before Richardson would publish his thoughts on numerical weather forecasting, he was already interested in meteorology when an iceberg found the *Titanic* in 1912. Soon after, also in 1912, Richardson filed British patent number 11125, for an "apparatus for warning a ship at sea of its nearness to large objects wholly or partly under water." Although echo sounding was far from a new idea, Richardson improved on existing methods, recognizing the advantages offered by shorter wavelengths — specifically, sounds with wavelengths of about three-quarters of an inch, which could penetrate through water to useful distances while still sending back echoes with reasonably high resolution.

Just like sonar, a sailor's lead line would find the bottom depth, but unlike sonar, a lead line — with the aid of a bit of wax applied to the bottom of the weight at the end of the line — could also bring a little piece of the seabed back aboard. A sailor examining the wax could see whether he was sailing above sand or mud or clay or rock. The depth and the bottom type were two pieces of information that might help a sailor with the ongoing, somewhat educated guessing about a vessel's location at sea.

My depth sounder tells me only the depth. But on the same panel, my chart plotter, linked to satellites, gives my latitude and longitude to within a few feet. My locational guesses are highly educated. They are, for a boat at sea, ridiculously accurate.

The wind swings to the east. We had hoped to anchor in Charlotte Harbor, but we fall off to the north. We aim for Venice. With this wind, we should reach the Venice breakwater before dusk.

With a hint of light on the horizon, we enter that period known as nautical twilight. That is, the horizon becomes visible, but stars still shine. It is — or was, years ago — the time of day for sextant-bearing navigators to measure the angle between the horizon and certain stars. Nautical twilight comes just before civil twilight, the period when objects at sea become visible, the period just before sunrise.

We sail toward that dimly lit horizon. I see a flashing light more or less on our course, but well ahead, a buoy in deep water.

Thirty minutes pass. Nautical twilight gives way to civil

twilight. Alongside *Rocinante*, I make out waves in the growing daylight. We are upon the flashing light, and I discern the shape of the buoy. It is a weather buoy, a bobbing tower standing on a round disk, twice the height of a person, keeping a lonely watch here, night and day, foul weather and fair, crewless, without a soul aboard, with no one staring at the waves, no one pacing on the bridge, no one longing for shore.

Today, the weather map is familiar. It is the most common map seen by newspaper readers, far outstripping maps showing political boundaries, war zones, crime locations, and electoral districts. To those who know the basics—the meaning of the lines of equal barometric pressure, or isobars, and the relevance of the lines and shapes showing moving fronts—the modern weather map captures far more than anyone could convey in the same space using mere words. It integrates information. The trained eye, the experienced user, understands at a glance not only the weather of the moment at any single location but also the weather of the moment across an entire geography. With the firing of a neuron or two, an intuitive short-term forecast materializes based on the realization that air masses move toward low pressure, arcing toward the right in the northern hemisphere, that certain systems follow an obvious track and will continue along that track, and that, for example, a cold front threatens to displace the currently sunny breeze with rain and strong gusts.

Most weather maps in newspapers, on television, and on the Internet are more properly known as synoptic weather maps. The word "synoptic" comes from the Greek word meaning something like "at the same time" or, more literally, "together view." It can also be interpreted to mean "the overall view," or, more simply, "the comprehensive view." When based on actual measurements, a synoptic map summarizing the previous day's weather is densely populated with odd lines, arrows, shapes, and numbers representing barometric pressure, wind speed, temperature, and cloud cover in various locations.

But the weather map is only apparently complicated. In contrast to the actual weather, it is wildly simple.

Today's synoptic weather maps cover areas hundreds of miles across. They illustrate weather systems. They show fronts and extratropical cyclones and other weather features that were only beginning to be understood a century ago.

What seems like the obvious starting point for forecasting—an understanding of current conditions, a snapshot of the weather—was at first seen only as a way to understand the workings of the atmosphere. For people who knew only about their own weather, who knew only what was happening outside their own windows, the compilation of information from scattered observers was informative. The plotting of that information on a map, an innovation that did not occur until the second half of the 1800s, must have been a revelation. Suddenly, at a glance, people could begin to see not just weather but weather systems.

The first maps came from the Victorian world of science, and that world understood the potential popular appeal. On April 1, 1875, the *Times* of London published the first newspaper weather map, showing the previous day's weather in Ireland, Great Britain, and parts of Norway, Denmark, Germany, and France. The weather in London had been "dull," with a temperature of forty-five. Hamburg's weather had been "thick" and cooler. Paris had experienced clear skies.

Despite popular appeal, the publication of weather maps in newspapers was sporadic until the early 1900s, with many papers not running maps at all and others running them daily for a time before losing interest. The problem was in part one of timeliness. Printers used individual engravings of letters that could be arranged as needed to form words, but pictures—including maps—were another matter. By the time a nineteenth-century newspaper could bring a weather map into print, the weather was old news.

Today, with streaming weather maps available on smartphones and televisions and computers, some of them animated, the challenges of publishing a map in the nineteenth century are hard to imagine. Take, for example, the production of a weather map in a British newspaper near the end of the nineteenth century. Observers at fifty stations telegraphed morning observations to London by ten o'clock. The Meteorological Office, receiving the data on a private wire, hand-drew two sets of maps, one for internal use and one, a simplified version, for the press. The press version was delivered by messenger to the Patent Type Founding Company. There an engraver traced wind arrows and lines of equal barometric pressure and equal temperature on a map. Templates and a drill were brought to bear, assisting the engraver's skilled eyes and hands. Completed, the engraving formed a mold that could be used to cast metal printing plates with the raised lines needed by printing presses. Messengers ran the plates to printers.

The director of the Meteorological Office complained about reports "taken once a day" and the delays in getting them to printers. He implied that a morning paper containing a map of the previous morning's weather was not entirely useful.

On the other side of the Atlantic, government weather offices scattered around the United States issued their own maps, sending them to local businesses, public offices, and schools. In 1891, more than a million maps, drawn by local weather offices and then printed, were sent out. By the turn of the century, the count increased to more than five million.

Congress balked at the cost. To save money, a single map, sent out by the Washington, DC, office, would be made available to newspapers. The papers would pick up the cost of printing. On March 1, 1910, the *Minneapolis Journal* printed what was referred to as the first "commercial weather map." Four months later, 65 morning papers in 45 cities were running the daily map. By 1912, 147 newspapers put the map in front of more than 3 million readers in 91 cities.

The distance between the meteorological offices in Washington and the major newspapers in other cities posed a challenge. Messengers could not cover the required distances quickly enough to move drawings and engravers' plates to printers. Instead, a coded grid system that could be transmitted by telegraph from Washington guided an artist in, say, New York, providing the artist with something akin to a paint-by-numbers template from which to build a map. Different artists served different cities and different newspapers. Costs ran high.

On first glance, these old weather maps look something like today's weather maps. They show outlines of states with temperatures, barometric pressures, and wind arrows. But on second glance, there is something missing. Before World War I, there were neither cold fronts nor warm fronts.

The lack of fronts did not reflect an intentional omission by mappers. Nor did it reflect the state of printing technologies. At the outbreak of World War I, meteorologists were simply not thinking in terms of weather fronts. Understanding weather in terms of fronts would require the work of Vilhelm Bjerknes, his students, and his son. The work that Bjerknes began led not only to the numerical forecasting that Lewis Fry Richardson would pursue on the battlefields of World War I but also to the appearance of weather fronts on maps, two very different but very productive approaches to understanding future winds.

The jetties at Venice, Florida, extend out into the Gulf, offering mariners a channel through the shallows, protected from waves on both sides by rocks piled on top of beach sand. With the jetties in sight by four in the afternoon, we lower *Rocinante's* sails and come in under power. We share the channel with bass boats and a motor yacht, a trawler.

Along the jetty, elderly men and women in folding chairs watch the boat traffic. They wave. They have no way of knowing that we have been offshore these past ten days. They have no way of knowing that we long for a hot shower and the peace of mind that comes with dock lines. They have no way of knowing that we set out for Mexico but have arrived here instead, off course, at a retirement mecca, home to people old enough to have met Vilhelm Bjerknes himself and to have lived through the changes wrought by his ideas.

They wave, and I wave back.

*Rocinante* sailed from Galveston with a green crew, untested at sea. We remain a green crew, raw recruits, but oddly, with ten days at sea under our belts, we are different people. We are sailors. Here in the harbor, we are interlopers.

Under power, we approach a small marina five minutes from the jetties. Maneuvering *Rocinante* in the marina's tight quarters is a negotiation. I want to ease gently against the dock and to stop there without bumping, but the boat herself, the spinning propeller, the currents, and the wind all have plans of their own. We approach slowly. My co-captain throws lines to dockhands. I use just enough reverse to bring us to a dead stop. Within seconds, lines fore and aft secure us. I kill the diesel. We have hit nothing, damaged nothing. Onlookers, ignorant of reality, might imagine that we know what we are doing.

Vilhelm Bjerknes was born in 1862, the same year Saxby published *Foretelling Weather: Being a Description of a Newly-*

*Discovered Lunar Weather-System,* the book that laid out forecasting methods based on the position of the moon. Three years later, Robert FitzRoy killed himself. Twenty years later, forecasting remained a matter of experience and intuition.

As a youth, Bjerknes ran experiments to verify his father's theoretical predictions about hydrodynamics. Son and father thought about the ether, the fluid medium that, they believed, allowed light and magnetism to move between planets and stars. Just as a boat feels the wake of another passing boat through the disturbance of the water's surface, a planet would feel the influence of another planet through the disturbance of the ether. In the greatest nineteenth-century minds, in a time before the acceptance of emptiness, in a time before Einstein brought his very special genius to bear, the existence of the ether was largely accepted. It was thought of as a rarefied gas of sorts, frictionless and mystical. But in fact it did not exist outside of the minds of scientists. And even there, this imaginary fluid served a purpose. It was this imaginary fluid that led Bjerknes to new understandings of fluid dynamics and, ultimately, of the atmosphere and its weather.

Like all scientists, Bjerknes was influenced by peers and predecessors. Among these was Hermann von Helmholtz, the famous German scientist, an authority on currents flowing in the ether. But Helmholtz did not limit his thinking to imaginary fluids. He was also interested in weather. He was not optimistic about theories explaining the atmosphere. He considered wind and the precipitation it carried to be the one natural phenomenon "that changes most deceptively, that impulsively and uncatchably evades any attempt to be caught under the bounds of law." In 1884, he summed up his view of weather by quoting what is today remembered as a children's song: *"Es regnet wenn es regnen will, es regnet seinen Lauf, und wenn's genug geregnet hat, so hört es wieder auf."*

That is, "It rains whenever it will please, it runs its course, the rain, and when it has rained long enough, it will stop once again."

The scientists wanted weather to behave. They wanted to understand the laws governing the movement of air. Without physical laws, weather could not make sense.

Bjerknes, at the turn of the century, was thinking about vortexes in fluids. He was thinking about spinning twisters and cyclones not in air, not in the atmosphere, but in the ether, in the perfect fluid that occurred everywhere but that could not be seen, the invisible stuff believed to allow the transmission of light and magnetism through empty space that had built his father's career and shaped his own life.

For a time, it was believed that vortexes could not form in this perfect fluid, or that if a vortex existed it would spin forever. The entire field was a playground for theorists, scientists with their heads in mathematical clouds. For scientists with their heads in actual clouds, for meteorologists, this abstract field was of little interest. For these scientists, working in real fluids, vortexes could form, and they did not spin forever. In the atmosphere, vortexes came and went. Some were a few feet across, and others covered entire regions. Some spun quickly, and others spun slowly. Some were benign, and others brought hardened sailors to their knees in desperate prayer. Atmospheric vortexes— whirlwinds and tornadoes and cyclones of all kinds—formed and disappeared, coming and going with confusing frequency and with no appearance of regularity.

But Bjerknes was not thinking about the atmosphere as he labored on what became known as his generalized circulation theorem. He was hoping to work out the mathematics behind electromagnetic phenomena in the ether. His work led him to see—in a purely mathematical sense—the formation of vortexes. These vortexes he saw should not have been there. They contradicted earlier work by well-known predecessors, men like

Helmholtz and Lord Kelvin. And yet there they were, mathematically existent, just as they existed in the atmosphere.

Bjerknes talked about his vortexes during a meeting of the Stockholm Physical Society in 1897 and again in 1898. On the second occasion, he referred not only to the ether but also to the atmosphere. He recognized that his work might be relevant in the real world.

The Norwegian's talk provoked other talk. Meteorologists were not especially interested in what might happen in an idealized fluid, but they were interested in what seemed to them to be the application of physics, the application of mathematics, in understanding the vagaries of the actual atmosphere. Here were ideas that went far beyond observation and bluster, ideas that complemented those of Ferrel, ideas that could advance meteorology beyond its lowly position as a descriptive science.

Meteorologists encouraged him. He wrote papers. He gave more talks. Following storms in 1902 and 1903, he developed his ideas further, eventually expressing his belief that his mathematical methods could forecast the weather.

In 1904, Bjerknes published a seven-page paper called "Das problem der Wettervorhersage, betrachtet vom Standpunkte der Mechanik und der Physik," or, in English, "The Problem of Weather Forecasting as Seen from the Standpoint of Mechanics and Physics." The paper did not announce so much the birth of modern forecasting as the expectation of that birth. Bjerknes, it might be said, announced the pregnancy that would bring modern forecasting into the world.

His paper was not peppered with numbers and mathematical symbols. In principle, what he suggested was straightforward. "If it is true," he began, "as any scientist believes, that subsequent states of the atmosphere develop from preceding ones according to physical laws, one will agree that the necessary and sufficient conditions for a rational solution of the problem of meteorological prediction are the following: 1. One has to know with sufficient accuracy the state of the atmosphere at a given time. 2. One has to know with sufficient accuracy the laws according to which one state of the atmosphere develops from another." In other words, one had to understand initial conditions, and one had to have working theories that explained what would come next.

The atmosphere's behavior, he went on, could be captured through seven calculations. There were three hydrodynamic equations of motion, one continuity equation, one equation of state, and two equations from the "fundamental laws of thermodynamics." Through these equations, one could, in principle, follow the movement of air and the water vapor it carried, tracking it as it changed temperature and as it traveled—as it blew— from regions of high pressure toward regions of low pressure.

He was not, in his writing, overconfident. "It must be admitted," he wrote, "that we may have overlooked important factors due to our incomplete knowledge." He speculated on the possibility, for example, of unknown cosmic effects and side effects "of an electrical and optical nature." He was under no illusion that mathematics would immediately address the challenges of forecasting. He knew that comparisons of forecasts with actual weather would be needed to improve the method. He knew that the method would require more data on current conditions— initial conditions—than were currently available. His paper includes phrases such as "fragmentary knowledge" and "instinct and visual judgment." And he knew that an absolute solution to the equations would be impossible and that even approximate mathematical solutions would be exceedingly difficult.

When it came to meteorology, he was not ready to entirely abandon qualitative judgment and experience. He was not ready to count on mathematics alone. He proposed the use of what he called "graphical and numerical methods." In other words, he proposed combining weather maps and mathematics, two different tools, each with strengths and weaknesses. The maps would start with the here and now and progress into the future in short increments, with the progression relying in part on judgment and in part on calculations. This method, he believed, would be "easily implementable." And this method led to many of the things taken for granted today, including the importance of fronts.

Later in life, he described his transition from theoretical physics into the realm of weather. "I entered," he wrote, "as an interloper into a science strange to me, meteorology." Bjerknes, an accidental but welcome guest in the world of moving air, brought with him the gift of objective science.

At the dock, I listen to the creaking of lines and the slapping of waves. I hear, for the first time in ten days, cars. And I hear, if I pay attention, air moving through the rigging, readily audible to attuned ears and alert minds.

A gull lands on the dock next to *Rocinante*. I watch it for a moment, and it watches me, cocking its head sideways, quizzically, possibly wondering if I am good for a handout. It hops toward me, springing along on yellow feet. It extends its wings, but not so much for flight as for balance. After a moment, losing interest, it hops to the edge of the dock, extends its wings farther, and steps off the edge, facing into what little wind blows. It drops a few inches, almost touching the water before gaining altitude.

In the third paragraph of the 1970s bestseller *Jonathan Livingston Seagull,* the bird for which the book is named maneuvers high above the sea with its wings curved. "The curve meant that he would fly slowly," wrote Richard Bach, "and now he slowed until the wind was a whisper in his face, until the ocean stood still beneath him."

Bach was, among other things, a keen observer of birds in the wind. So were Orville and Wilbur Wright, whose early experiments with flight coincided in time with Vilhelm Bjerknes's early thoughts about forecasting. In 1900, the brothers searched for a place to test their flying machines. Knowing they wanted a location where winds blew reliably at fifteen miles per hour, they searched government weather records. They sent out query letters. A response came back from a man named William J. Tate. The letter described a place with "a stretch of sandy land one mile by five with a bare hill in center 80 feet high, not a tree or bush anywhere to break the evenness of the wind current." That wind current, Tate wrote, blew steadily, "generally from ten to twenty miles velocity per hour." From their home in Ohio, with help from Tate and the US government, the Wright brothers found Kitty Hawk, North Carolina, seven hundred miles away.

Once at Kitty Hawk, Wilbur and Orville watched birds. Specifically, they watched the wings of birds, the ways in which birds angled their wings. "The buzzard," wrote Orville in his notebook, "which uses the dihedral angle finds greater difficulty to maintain equilibrium in strong winds than eagles and hawks which hold their wings level."

The neighbors in the then isolated community of Kitty Hawk watched the brothers watching birds. "We couldn't help thinking they were just a pair of poor nuts," wrote one neighbor. "They'd stand on the beach for hours at a time just looking at the gulls flying, soaring, dipping." According to the neighbor, the brothers moved their arms in imitation of wings.

That first year at Kitty Hawk, they played with a glider. In the beginning, the glider flew unmanned, more or less as a kite. In October 1900, Wilbur made his first manned glider flight.

*The Wright brothers' 1900 glider on an unmanned flight at Kitty Hawk. Three years later, the brothers would make their first powered flight. (Image from the Library of Congress and Wikimedia Commons)*

Back in Dayton, in the bicycle shop where they earned their living, the brothers built a wind tunnel. It was nothing more than a six-foot-long open-ended box with a fan mounted at one end. Inside, they tested different designs of tiny wings made from old hacksaw blades. For weeks they tested different designs, thirty-eight in all, with wind speeds in their small tunnel reaching twenty-seven miles per hour.

In 1903, they returned to Kitty Hawk, but this time with an engine. That year, while Bjerknes was talking about the science of moving air, the Wright brothers were flying through it.

Their first powered flight lasted twelve seconds and carried Orville Wright 120 feet, about half the length of a Boeing 747. All the early flights were mercifully short—less than a minute

*The Wright brothers' wind tunnel. The original photograph was taken around 1901 but has been reproduced by NASA and others.*

each—and close to the ground, but they were the start of a rapid evolution in powered flight. In 1909, an airplane crossed the English Channel. In 1911, pilot Calbraith Perry Rodgers hopscotched across America, spending more than eighty hours aloft dodging bad weather during forty-nine days of travel from Long Island, New York, to Pasadena, California. In 1933, just thirty years after the first powered flights at Kitty Hawk, a biplane flew above the peak of Mount Everest, the highest mountain on earth.

Whatever the demand for forecasting had been in the days leading up to the twentieth century, the sudden appearance of aviation upped the ante. As early as 1918, just fifteen years after the Wright brothers' first powered flight and a year before the first transatlantic airplane flight, the US Weather Bureau posted its first aviation weather forecast, supporting military flights as well as mail flights that were already making the run from New York to Chicago.

In 1926, Congress passed the Air Commerce Act, which, among other things, directed the Weather Bureau to provide forecasts and warnings "to promote the safety and efficiency of air navigation in the United States." The limited early aviation forecasts were based almost entirely on data collected from the ground. In 1931, the Weather Bureau began regular observations at altitude using airplanes in Chicago, Cleveland, Dallas, and Omaha. The bulk of the effort went toward describing current conditions rather than future conditions, but at least pilots would know what winds were already out there.

The gull, or another one that looks just like it, lands again, this time next to the cockpit on *Rocinante* herself. And again, it looks at me with a tilted head. Its expression, if gulls can have expressions, is inquisitive. When I approach, the gull stretches its wings, catches the breeze, and hops into the empty air.

Later, I read of the death of Calbraith Perry Rodgers, the first man to fly from coast to coast across the United States. In 1912, less than a year after his historic flight, a flock of gulls collided with his plane over California, the plane crashed, and Rodgers died.

Bjerknes did not shy away from sharing his ideas. In 1904, the year of his seminal seven-page technical paper on forecasting, one year after the Wright brothers' first powered flight, Bjerknes wrote "Weather Predictions and the Prospect for Their Improvement" for the Norwegian newspaper *Aftenposten*. The newspaper article was twice the length of his technical article. In it, he translated his ideas for the public. Put another way, he promoted his ideas. He went in like a man selling anything new, introducing himself, convincing readers of the importance of what he had to share, offering first the familiar, then moving to explanations of the unfamiliar, and, toward the end, assuring readers that nothing would stand in the way of forecasting.

He let readers know that he was new to weather forecasting. He referred to other scientists. "Through them," he wrote, "I have been brought to meteorology, without my knowledge and my will, I might nearly say."

He captured the importance of his work by suggesting, in his first line, that humans were driven by a desire to understand what would come next. "After closer examination," he wrote, "the thought of the future is hiding behind our struggle for knowledge." Students went to school to better understand how their decisions would affect their future and the future of society. Doctors, he said, were in the business of predicting future health. Businessmen succeeded by understanding how investments would eventually lead to returns. Even historians shared the goal, working to understand the "continuous development from the past, through the present, and towards the future."

Future weather—the ability to forecast the weather of tomorrow and the next day and the day after that—would be important to seamen, as readers in the maritime nation of Norway would know. But it would also be important to agriculture. Bjerknes mentioned the late Robert FitzRoy, calling him "the first pioneer" of forecasting.

He acknowledged what his readers already knew. Forecasting, at the beginning of the twentieth century, was not perfect. There were many weather stations, enough to allow the drawing of weather maps. The modern telegraph permitted communication between stations. Progress had been made, but it had been slow. In some places, including parts of Norway, forecasters had no real hope of predicting future events. "The feeling of the distance between what has been obtained and what ought to be obtained," he wrote, "seems also to have exerted a paralyzing effect on more than one meteorologist, who has gone tired under the stressing work with the eternally changing temper of the weather."

The cure for this paralysis, he suggested, was "new weapons." He called for more weather stations. He called for aerial observations, using not only the recent invention of instrumented balloons but also instrumented kites, some of which were already flying at altitudes higher than two miles. Most important, he called for the application of the laws of physics and mechanics. He called for the application of mathematics.

A year later, Bjerknes presented his ideas in New York and Washington. "The program excited the interest of the Americans," he later wrote, "perhaps because of its boldness and colossal dimensions." Recognizing the importance of his ideas, the still young Carnegie Institution provided support for his meteorological research in the form of generous funding year after year. "From 1906," Bjerknes wrote, "I had, thanks to an annual grant, the possibility of taking on one or more personal scientific assistants. Thus my fate was sealed."

His work led him to Germany. There, his ideas matured. "Concerning scientific work," he wrote to a friend, "the final year in Leipzig has been extremely useful. For the first time, we have made headway with meteorological prognosis based on dynamical principles. How much practical significance this might have, it is still too soon to say." In other words, he advanced the marriage of observation and theory, a marriage that produced forecasts, or, as he called them, "prognoses."

As Bjerknes labored, European conflicts grew into World War I and rendered life and work in Germany untenable. After losing students to the war effort, Bjerknes returned to Norway. In Bergen, he launched what would become known as the Bergen School. He lived downstairs, while upstairs he, his son, and his students and colleagues labored over maps and charts, thinking about the weather.

Before the end of his first year back in Norway, he sent a letter to a friend. "We are really getting on with the prognosis problem, using dynamical principles," he wrote. By now, "dynamical principles" were understood to be synonymous with the underlying theory that treated the atmosphere as one would treat any other fluid.

"Up to a certain point," Bjerknes continued, "it is going well, and strangely enough it is my old circulation theorem which seems to offer most, so far at any rate. But other paths appear to be opening up too—competition is healthy after all. There is still a long way to anything practical, but in any case it feels satisfying when it turns out that atmospheric phenomena develop according to the laws of nature."

More so than his predecessors, men like Helmholtz, he saw the laws of nature restricting the atmosphere's playfulness. The atmosphere did, after all, follow the laws of nature. But the important points in his letter—points that may have escaped Bjerknes himself at the time—were about "other paths" and the "practical."

When he wrote the letter, Norwegians were feeling the hunger pangs of war. In 1916, the year before he returned home, Norway produced less than half the grain needed by its citizens. The wartime challenges of shipping drove the price of imported food skyward. In 1917, the harvest fell well short of expectations.

On February 13, 1918, Bjerknes came across a newspaper story describing a Swedish weather forecasting service that offered forecasts to farmers via telephone. In response to the article, a Norwegian official claimed that such a scheme would not work in Norway. Bjerknes disagreed. Not only could it work, but it was the patriotic thing to do. Bjerknes knew that short-term forecasts would help farmers, and helping farmers would help the country. He had a conversation with Norway's prime minister. The nation's legislature, the Storting, offered him a grant of one hundred thousand kroner to support forecasts for farmers.

"Life is fateful," Bjerknes wrote to his friend. "Now I have suddenly become a practical meteorologist. We shall try to do all we can in order to provide weather forecasts for farming."

We wander the docks talking to other boaters. Most are weekenders. Two, in motor yachts, are "Loopers," meaning they came down from Chicago via the Illinois River, the Mississippi River, the Tennessee River, the Tennessee-Tombigbee Waterway, and the Tombigbee River to Alabama's Mobile Bay, then along the Florida coast to Venice. From here they will press on, crossing Florida on Lake Okeechobee before turning north, eventually following canals into the Great Lakes and closing the loop back in Chicago. They have their own association, the America's Great Loop Cruisers' Association.

Loopers do not, with rare exceptions, sail. The route goes under low bridges and over shallow shoals. Masts and deep keels are not welcome. Nevertheless, they worry about the wind. A Looper's boat, if caught without the stabilizing benefit of a sail in the few stretches of open water that must be crossed, would wallow dreadfully in wind-driven waves of more than a few feet. The crew, accustomed to rivers and lakes, could get seasick. The inevitable dirt at the bottom of the fuel tank might be suspended, clogging the fuel filter, stalling the engine, and adding to the general misery of all hands. Loopers, as much as sailors, have to watch the wind.

We go to the marina's tavern for a fresh salad. A waitress takes our order, and a high-definition television displays the weather. The woman on the screen, in heels and a green dress, her hair anything but windblown, points at a map of the nation. Satellite imagery overlays an outline of the states. The imagery flashes forward. She points to lines indicating fronts. She smiles. She talks fast.

Fronts were not formally described until well into the twentieth century, but the knowledge of moving parcels of air predate their formal description. They can, after all, be observed directly, as storms moving across the landscape. As early as 1828, the year young Robert FitzRoy was appointed captain of the *Beagle*, movements of cold and warm air currents were described, and by 1841 a drawing of two air parcels meeting was published, with one parcel labeled "north-west current" and the other labeled "south-west current."

Today, people talk about fronts in casual conversations. They are prominent features of most weather maps. The weather forecaster on the screen talks about them authoritatively. She warns viewers of the cold front that is coming our way, bringing with it wind and rain. What she means, though, is that the cold front will encounter warm air. The warm air, relative to the cold air, is buoyant. The warm air will float above the cold air. As it gains altitude, the warm air will cool. When the warm air cools, its water vapor will condense. The cold front will induce rain out of the warm front, but it does not bring the rain with it.

Raquel Welch once forecast the weather for television viewers. So did Gilda Radner, before she joined *Saturday Night Live* and played Roseanne Roseannadanna. And Diane Sawyer, later with *Good Morning America* and ABC's *World News*. And Pat Sajak, later with *Wheel of Fortune*. And Bob Iger, later chief executive officer of the Walt Disney Company. And David Letterman, who included fictional cities in his forecasts, who once predicted hail the size of canned hams, and who once congratulated a storm that had been granted hurricane status.

Television weather started in the early 1940s. New York City's first weathercast, on October 14, 1941, starred Wooly Lamb, an animated character with a telescope who talked in rhymes. There were gimmicks, then as now, tricks to hold viewers. But from the beginning, some forecasters saw television as an opportunity to educate viewers about the workings of the atmosphere. Viewers knew nothing of cold fronts and warm fronts and low-pressure systems. Meteorological jargon now in common use was unfathomable to early viewers. Forecasters sketched weather maps in front of the cameras, taking science into living rooms in black and white.

The American Meteorological Society rallied behind serious forecasters, offering its Seal of Approval. From a 1955 article by a physics professor writing for *TV Guide:* "We think the weather should be discussed with dignity. Dignity, not dullness. We think many TV 'weathermen' make a caricature of what is essentially a serious and scientific occupation."

Twenty-seven years later, The Weather Channel debuted. Weather documentaries aired on a regular basis. Another thirteen years passed, and the Weather Underground, born from an academic program, offered global real-time weather data and interpretation through the Internet. Other online weather portals proliferated. What had once required couriers and block printing could now be accessed from home computers and cell phones.

From our table we watch the woman in the green dress. Meteorologists can be trained in mathematical meteorology or broadcast meteorology or something in between. I cannot tell whether she is a trained meteorologist of any kind, or whether she simply learned on the job. I do not know whether she has the approval of the American Meteorological Society. I do know that she wants viewers to understand that a cold front may be on the way. It may die before it reaches Venice, or it may not. She is cheerful about her uncertainty, but also dignified.

She says nothing about where the forecast came from, about data streams from buoys and satellites, about numerical models. She is probably doing little more than repackaging information provided by the National Weather Service. She mentions neither

Espy nor Redfield, Bjerknes nor Richardson. And to be fair to her, the public may not be predisposed to those details, even if I am.

More and more, Bjerknes straddled two worlds, the theoretical and the real, the numerical and the experienced, the academic and the practical. But the practical drew most of his attention. It was this straddling of two worlds that enabled Bjerknes, with his students and assistants, to transform predictive meteorology. They combined maps and mathematics, applying graphical methods to solve otherwise unsolvable problems, tempering their output with experience.

In 1918, young Jacob Bjerknes, son of Vilhelm Bjerknes, began to develop ideas about large parcels of air, about how they moved and interacted, about how a cold parcel might meet a warm parcel. At the Bergen School, immersed as it was in the news and hardships of the Great War and its battlefronts, knowing of the troops bogged down in trenches, Jacob Bjerknes and his colleagues eventually called the leading edges of the moving parcels "fronts." As anyone today would recognize, where a cold front meets a warm front, the weather often turns violent.

*Chapter 6*

# THE MODEL

Vilhelm Bjerknes lived until 1951. In 1944, the octogenarian Bjerknes described his ongoing duties at a research center heated by wood. "We looked in vain for a new stove boy," he wrote. "I was the one who had free hands to do this work. With my old and slow-working brain I regarded myself as quite unfit to compete with the young meteorologists or assistants in the express-work at the weather maps. But I tried to keep the fire going, both spiritually and materially."

The fire he started and maintained bridged the gap between those who merely observed the weather and those who merely theorized about it. Systematic observations of initial conditions and an understanding of the physical behavior of the atmosphere pushed forecasting far beyond the intuitive methods of men like Robert FitzRoy.

Bjerknes was born into a world that took Stephen Martin Saxby seriously—Saxby, whose book explained how the position of the moon foretold of coming storms. Before Bjerknes died at the age of eighty-nine, he would see his students and collaborators move the understanding of the earth's atmosphere to new levels. He would also see the forecasting methods of the Bergen School slowly take hold, spurred on in part by wartime demands, in part by the rapid evolution of aviation after World War I, and in part by the short-lived but tragic meteorological event known as the Dust Bowl. He would see the routine application of his graphical methods applied in forecasts. Bjerknes, the reluctant interloper in the world of meteorology, would see Lewis Fry Richardson complete an entirely mathematical forecast using an approach that Bjerknes himself had suggested but never tried. And toward the end of his life, he would see mathematical methods programmed into ENIAC, the Electronic Numerical Integrator and Computer, one of the world's early electronic computers.

Like Bjerknes, Lewis Fry Richardson might be described as an interloper in the world of meteorology. He was the youngest of seven children, born into a Quaker family in England in 1881. His career neither began nor ended with weather.

Richardson's earliest work of note was undertaken when he was employed by a peat-mining company called National Peat Industries. At that time, peat was commonly mined from bogs as a source of fuel, an intermediary between wood and coal that could be dried and then burned in household fireplaces and industrial stoves. The company, according to its 1905 prospectus, emphasized the importance of science and engineering, especially in the realm of drying the peat.

Richardson, in his midtwenties, invoked a well-known equation to understand the movement of water through peat bogs, but then modified the equation to account for the irregular geometries of the real world, a world that cannot always be represented by straight lines and neat circles and squares. He was working with differential equations, many of which could not be solved. It was not that Richardson could not solve these equations, but that they were, by their nature, unsolvable. He applied a method that offered approximate solutions. The method, known as the finite difference method, involved cutting mathematical corners to get an answer that, though not perfect, was good enough. This experience would prove very useful when he turned his attention to weather forecasting.

Like many young scientists today, Richardson held a series of temporary jobs. In addition to National Peat Industries, he worked for the Sunbeam Lamp Company and various government laboratories. In 1913, he landed at the Eskdalemuir Observatory in Scotland. The observatory recorded seismic events and measured magnetic fields, but it also gathered meteorological data. Richardson found himself tasked with improving the theoretical understanding of weather.

While at the observatory, he began working on his book *Weather Prediction by Numerical Process*. The book starts with a description of a system for indexing weather maps. The idea was simple: the past behavior of the atmosphere would be used to predict the future behavior of the atmosphere. With the aid of an index of weather maps, one would be able to compare today's weather map with previously compiled maps. Finding a match in the index, one could then flip forward a day or two, using the past sequence to predict the future sequence. "It would be difficult," Richardson wrote, "to imagine anything more immediately practical."

But fewer than 150 words into his book, at the beginning of the second paragraph, he dismisses the method. Although it would be practical, in the sense that detailed weather records covering decades were available, the method, he knew, would not work. One would not use such a method to predict the future position of the planets, and one should not use such a method to forecast the weather. Instead, he promises to present a method "founded upon the differential equations, and not upon the partial recurrences of phenomena in their ensemble."

Richardson's first mention of Bjerknes appears in the fifth paragraph, near the top of the second page. "The extensive

AND SOON I HEARD A ROARING WIND

researches of V. Bjerknes and his School are pervaded by the idea of using the differential equations for all that they are worth," he wrote. Yet Bjerknes himself, justifiably believing the mathematics to be intractable, used graphical methods. Richardson acknowledges this difference, writing, "Whereas Prof. Bjerknes mostly employs graphs, I have thought it better to proceed by way of numerical tables."

Richardson was not criticizing Bjerknes. He was not suggesting that Bjerknes should have proceeded with numerical tables. He was merely following through on something that Bjerknes believed to be plausible in principle but exceedingly difficult or impossible in reality. But unlike Bjerknes, Richardson was armed with an intimate knowledge of the mathematical shortcuts that might render the apparently undoable doable.

Richardson had been thinking about his approach and his book as early as 1911 and had given it what he called "serious attention" since 1913, when he was encouraged by the director of the Meteorological Office, a successor to the long-dead Robert FitzRoy. "The fundamental idea," Richardson wrote, "is that atmospheric pressures, velocities, etc. should be expressed as numbers, and should be tabulated at certain latitudes, longitudes and heights, so as to give a general account of the state of the atmosphere at any instant, over an extended region, up to a height of say 20 kilometers." In other words, he suggested laying out a three-dimensional grid, populating the grid at time zero with data from observations, and then using mathematics to calculate future conditions as air moved from one cell to another. He suggested, in effect, turning the atmosphere into a three-dimensional chessboard, with air parcels as pieces, their movements governed by complicated but rigid rules.

On a global scale, each square near the equator would be about 200 miles from north to south and about 130 miles from east to west. The dimensions were chosen in part to accommo-

136

date available data in a world sparsely populated by weather stations, but they were partly chosen as a matter of convenience. Smaller dimensions would mean more cells, and more cells would mean more math.

Vertically, he divided the atmosphere into five layers. "Ten layers would give four times the accuracy obtained with these five layers," he wrote, but again, there was the matter of convenience.

In the real world, where weather changes from one instant to the next and from one spot to the next, where clouds move constantly and wind speeds and directions ceaselessly change, the mathematics that Richardson envisioned would never represent reality. He had to simplify the real world. In the language of mathematicians, the future of the real world could only be seen through the application of unsolvable differential equations, but the future of his simplified world could be seen through the application of solvable finite difference methods. Instead of looking at the world as it truly is—a world in perpetual and unresting flux—he chopped time and space into mathematically manageable chunks, as though change occurred in increments, one increment clearly separate from the next. He was applying the very same methods he had applied as a younger man employed by the peat industry, where the irregular geometries of peat bogs had to be simplified before he could solve problems related to drainage. With weather as with peat, the mathematical solutions would not be perfect, but they might be good enough.

In Richardson's simplified world, in his three-dimensional chessboard, the weather across a cell would be constant and consistent for six hours, after which it would change abruptly. Each cell would affect each neighboring cell.

Richardson knew, of course, that the real world was not composed of discrete cells. He knew the feeling of constantly changing winds and of gradually but constantly changing temperatures. Taking this reality to an extreme, he knew that the very idea of wind speed as a meaningful concept would not withstand challenge. In a paper about mathematical methods, he considered the problem of measuring the speed of a fluid that in one sense might be moving as a whole, but that in another sense was full of eddies, gusts, and calms. "Does the wind possess velocity?" he asked. "The question, at first sight foolish, improves on acquaintance." His point was that wind—moving air—is a composite of moving particles, of individual molecules each finding their own way, each constantly changing not only its speed but also its direction, even as the air as a whole undeniably moves with both a discernible direction and speed. Although he knew that the Robinson anemometer did a reasonable job of averaging the movements of those uncountable molecules of air, of smoothing the appearance of what is by nature a very complex process, he also knew wind for what it was.

"We are not concerned," he wrote in *Weather Prediction by Numerical Process*, "to know all about the weather, nor even to trace the entangled detail of the path of every air-particle. A judicious selection is necessary for our peace of mind." Richardson was something like a historian who cannot dwell on the daily lives of every citizen, but instead must look for the important events, the key factors that shape the world. In his equations, Richardson could not look at every aspect of the weather. He could not allow himself to be bogged down in the endless details. He knew that a still night on the ground might be a windy night aloft, above the treetops. He knew that smaller cells and shorter increments would offer more realistic results, but somehow he had to choose those aspects of the weather that mattered most. Tracking hourly movements would have improved his results but made the already challenging math less manageable. Twice-daily movements would have cut his labor in half, but his results would have suffered. "Also," he wrote, "the errors increase with the number of steps."

He had to select the important facets of today's weather, those that would have the greatest influence on tomorrow's weather, those that would allow him, with his mathematics, to see the future. He knew that calculations undertaken with pencil and paper would contain mistakes, and that the design itself, with its large cells and long time steps, would lead to errors. And he knew that neither the mathematics nor the initial conditions would be entirely right.

The work was fraught with uncertainty. He wrote that some of his tabulated values reminded him "of the stories which are 'founded on fact,'" possibly referring to books such as *Robinson Crusoe* and *Moby-Dick*. One of his footnotes says, "Probably, but I have no definite information."

Chapters have titles such as "The Fundamental Equations" and "Finding the Vertical Velocity." The contents of the chapters were not exactly easy reading. "Now to represent the initial observations of pressure," he wrote, "we are at liberty to write down any arbitrary set of numbers, at the points of the map where $p_G$ is required, only with this qualification: that if the assumed pressure gradients be unnaturally steep, the consequent changes will be perplexingly violent."

Or this gem: "The advantage of the chessboard pattern is now seen to be that the time-rates are given at the points at which the variables are initially tabulated."

And this: "The road to a fuller knowledge of the variations of viscosity appears to lie through a study of the diffusion of eddies."

The necessary calculations were repetitive and tedious. Richardson turned to his wife, an accomplished mathematician herself, for assistance.

While he labored on the first chapters of his book, assassins shot Archduke Franz Ferdinand of Austria and his wife, Sophie, the Duchess of Hohenberg. World War I descended on Europe. Richardson, a Quaker and a committed pacifist, would not fight, but he volunteered for ambulance duty. He took his unfinished manuscript to the front, and there he worked on a concrete example.

"Let us now illustrate and test the proposals of the foregoing chapters," Richardson wrote, "by applying them in a definite case supplied by Nature and measured in one of the most complete sets of observations on record." He wanted to test his methods. He planned to apply them to a date in the past, to generate a forecast that he could then compare with historical reality.

He started with observations from the middle of Europe collected on May 20, 1910. The measurements came from a coordinated effort that involved both work on the ground and work aloft, with the aid of balloons. Bjerknes himself had coordinated the observers and compiled the data. "This region and instant," Richardson wrote, "were chosen because the observations form the most complete set known to me at the time of writing."

But the data were far from perfect. "At some points there is large uncertainty," he wrote. At other points, data were entirely missing, forcing him, in essence, to guess.

For each cell and each interval, he started with four variables that were entirely independent of the effects of weather: time, height, latitude, and longitude. To these he added seven variables that represented the weather and that changed as the weather changed: three for wind speed in different directions, one for the density of the air, one for the amount of water held in the air, one for temperature, and one for pressure.

"If an eighth dependent variable had been taken," Richardson wrote, "it might perhaps have specified the amount of dust in the air." In other words, he knew that potentially important factors would be ignored.

The war was in full swing. Richardson, in his role as ambulance driver, made his calculations while stationed very close to the front lines. He worked between ambulance runs with a slide rule and pencil and numerical tables. He dreamed of his mathematical forecast. Later in life, Richardson wrote a few lines about his time as an ambulance driver: "I was a bad motor-driver because at times I saw my dream instead of the traffic."

Men working with him near the front lines remarked on his often quizzical expression. They knew him as someone who might become suddenly detached, distracted by a new idea in the middle of a conversation. Most of his colleagues in Section Sanitaire Anglaise Treize, his Red Cross ambulance unit in France, were younger than he was, and they referred to him as "the Professor" or, sometimes, "the Prophet."

An account written in 1919, called "L'automobile dans la guerre," describes the ambulance drivers as a "veritable elite" whose actions were cited after every engagement. "The fact is," the account continues, "that the ambulance drivers, who theoretically were only supposed to go as far as the advance posts of the Division Stretcher-Bearer Corps, went all the way up to the aid stations to pick up their wounded, across terrain mined by the enemy, through intense rifle fire and waves of gas which they had to cross in the open. Add to this the darkness, the cries of the wounded for whom each bump in the road meant torture, the gas mask which made it so terribly difficult to drive!"

The duty was neither glamorous nor conducive to producing a numerical forecast. "Each day we went on duty to the dugout," recalled one of Richardson's comrades, "we stopped on our way to pick up some wood such as laths, timber frames, etc. from the ruined villages en route, to keep the fire burning. From here we went to the Front to pick up the wounded when a call came through. We slept in the back portion of the dugouts, in full clothing so as to be ready at a moment's notice. These dugouts were very stuffy at night and I used to waken up with a headache. Also, we had rats running around, which very often ran over us as we lay down."

During the Third Battle of Champagne in April 1917, Richardson's manuscript was sent to the rear, apparently to protect it, but the manuscript was lost. Months later it was found under what Richardson described as "a heap of coal." He provided no further explanation, leaving readers to imagine that finding a lost manuscript capable of changing the way the world viewed weather buried under a heap of coal was perhaps barely noteworthy.

During major battles, ambulance drivers would have been too busy and too exhausted for mathematics. Richardson's calculations occurred between enemy engagements and during brief breaks from the front.

"My office," he wrote, "was a heap of hay in a cold rest billet." There in the rest billet, he worked through the math. In fact, he worked through the math twice, checking and correcting his results. Satisfied with his calculations, he compared his numerical forecast with what had actually transpired. His results did not match the reality.

"The rate of rise of surface pressure," Richardson wrote, "is found on form XIII as 145 millibars in 6 hours, whereas observations show that the barometer was nearly steady." For perspective, the normal range of atmospheric pressures at sea level is about 70 millibars, from a low of 980 to a high of 1,050. Richardson's numerical forecast was not only wrong, it was wildly and impossibly wrong. A guess informed by no more than the possible range of pressure changes at sea level would have been more accurate.

Richardson called the mismatch between his mathematical forecast and the reality on the ground a "glaring error" and later referred to "striking errors in the forecast." Despite this, he expressed neither disappointment nor frustration. A conscientious objector surrounded by the wounded and the dead, he appears to have accepted the mismatch with a sense of perspective. The glaring error was a matter of interest rather than personal failure. It did not prevent him from publishing his book, and it did not prevent him from dreaming of a time when the calculations could be run faster than the weather itself unfolded, of a time when numerical forecasting could predict future weather, when the graphical methods of Bjerknes could be replaced by what men like Richardson would consider the elegance of numbers.

My co-captain and I awaken early and listen to the mechanical voice of the government's marine weather station. The wind will blow at five to fifteen knots from the east and northeast for the forecastable future. The government promises a wind that will take *Rocinante* and her crew of two straight to Mexico. We agree to set sail at noon, giving us time for a walk on the beach.

We find seashells mixed with sand, both tossed ashore by waves, the waves themselves driven by wind. The beach seems narrow, as Gulf beaches go, with windblown dune plants a stone's throw from the waves. The wind blows above the plants, but in the shade of the foliage near the ground, the air lies still, the wind intercepted by leaves and stems, its energy lost to friction. Even more so than the morning sea breeze, air movement at this scale would have been entirely invisible to Richardson's model. As Richardson wrote, "We are not concerned to know all about the weather, nor even to trace the entangled detail of the path of every air-particle."

A stone's throw beyond the edge of the dune plants, condo-miniums stand at attention. Their windows face the Gulf, wel-coming the sea breeze.

A wind just over five knots—to Beaufort a light breeze—can move sand. Individual grains begin to vibrate, and then to roll, and then to bounce. As they bounce, they catch more wind. They leap. In the jargon of the trade, they "saltate."

If an individual grain rolls and bounces and saltates far enough, it might reach the dune plants. It might stall in the still air under the plants. If the sand grain rests there long enough, it might be gripped by roots.

Some plants depend on nutrients from droplets of seawater carried ashore by the wind. Others cannot tolerate the salt and retreat farther inland. As we walk, we see morning glory and sea oats and panic grass and sea rocket, each finding a place to sur-vive along the gradient of salt and shifting sands and moisture.

Individual dune plants, loners, stand on mounds of sand. Some send out rhizomes, creeping roots with stems and leaves at intervals, capturing more sand. Others grow in dense thick-ets, each benefiting from the others while also competing for space.

Collectively, the wind conspires with the plants to build land. An upward step marks the boundary between sandy beach and dune plants.

We want to be under way by noon. We turn to walk back toward *Rocinante*.

While Richardson worked on his calculations and wrote his book, Bjerknes and his colleagues at the Bergen School advanced their methods and their thinking, but they did so, for the most part, conceptually. Bjerknes might be the father of numerical forecasting, but, if so, he was to some extent detached from his offspring. The men at the Bergen School were making sense of moving air, but they were not moving toward numerical forecasting. They focused on what became known as air mass analysis. They viewed the atmosphere as packets of air. The packets—the air masses—interacted at fronts.

Warm fronts and cold fronts—discontinuities in the atmosphere—presented a problem for Richardson's weather models. "It has been the custom," Richardson wrote, "to regard line-squalls and other marked discontinuities as curious exceptions to the otherwise smoothly gradated distribution of the atmosphere." He was concerned that his mathematical approach might be stymied by the abrupt changes that occur at fronts. Bjerknes and his colleagues, according to Richardson, believed that "discontinuities are the vital organs supplying the energy to cyclones." And if they were the vital organs of cyclones, they were important to forecasting.

Although Richardson knew that his methods could not easily deal with fronts, he believed that knowledge of fronts alone could not forecast weather. "It is not to be expected," he wrote, "that a knowledge of the position and motion of surfaces of discontinuity will prove to be sufficient for forecasting, any more than 'vital organs' alone would suffice to keep an animal alive." He believed that his numerical methods would work up to the edge of the front, but in crossing the front the forecaster would have to improvise. Even in Richardson's view, mathematics alone were not enough. The forecaster would be forced to turn to intuition and experience. The forecaster had to accept subjective decisions where fronts were concerned.

Bjerknes himself, along with his colleagues and students at the Bergen School, believed in principle in a mathematical atmosphere. They knew that the movement of air would follow the rules of physics. They also knew that their graphical methods would not, when artfully applied, predict pressures that do not actually occur, whereas the mathematics applied by Richardson would. And they knew that their graphical methods could keep up with the reality of weather, whereas the mathematics applied by Richardson could not. Richardson's work was of interest, but it was not practical.

The graphical methods pioneered by Bjerknes, the Bergen School ideas, well developed by the end of the war and in retrospect very straightforward, were not immediately embraced. Fronts were not plotted in the British Daily Weather Report until 1933. Much of the funding for the Bergen School came from the United States, via the Carnegie Institution, but at the US Weather Bureau the ideas of the Bergen School got the same cold shoulder they received in Britain. The United States was not ready for Norwegian forecasting. Air mass analysis was not used by the Weather Bureau before 1934, and fronts did not appear on many of the bureau's maps before 1936. And if the mainstream weather forecasters were not ready for the Bergen School's graphical methods, they were certainly not ready for Richardson's mathematics.

Back on board, I still feel nothing of the forecasted five to fifteen knots from the east and northeast. When I face the sea, I feel five knots in my face—five knots from the west, from the direction we would like to sail. But it is only the gentle sea breeze, cooler air from above the Gulf blowing in to replace the air above the land, air warmed quickly under the morning sun, air sent aloft by the heat of daylight on the ground in compliance with Archimedes, its speed dictated by Euler. In the mathematical cells used by Richardson, winds at this scale, mere morning breezes, might be altogether invisible.

I remember something from a book that I keep on a shelf in

the galley. I should prepare for sea, but instead I pull the book down from our shipboard library. Somewhere in the book is a page copied from a ship's log. It takes me a moment to find it. The copy contains the captain's barely legible handwriting. In the middle of the page, a small but pretty watercolor shows the captain's own ship, a three-masted square-rigged vessel, all sails full, moving across waves that suggest a wind of twenty-five or thirty knots, somewhere between a strong breeze and a near gale. The ship sailed in 1790, before Beaufort's designations were in place, before FitzRoy, before the era that would revolutionize meteorology and forecasting. And yet the ship sailed. The log puts the ship three hundred miles from Cape Blanco, well offshore of Mauritania and what is today the Moroccan-controlled portion of the Western Sahara.

With a magnifying glass and significant concentration, I make out the line that drew me back to this book. "Sand is so loose and small," wrote the captain, "that a fresh breeze coming from the northeast sweeps it off and carries it to sea. When we first experienced it we were nearly 300 miles from the Cape so that the size of the particles of which it is composed and the quantity carried away may perhaps nearly be conceived."

This is the sort of wind that would have been visible to Richardson's model. The details might not be important, but a wind sweeping sand three hundred miles out to sea is a wind of note.

I put the book away and check the oil in *Rocinante's* engine. It is almost time to get under way.

In the early 1930s, the US Weather Bureau willfully ignored the concepts of moving air masses and fronts, along with the mathematics of wind, but the moving air masses and fronts, along with the mathematics of wind, did not ignore the United States.

Drought settled in across the middle of the continent. Winds blew. Fronts swept through. Particles of soil vibrated, then bounced, then leaped, then flew. Particles of soil saltated with a vengeance.

In 1934, more than twelve million pounds of soil from the Great Plains settled in Chicago. Other soil from the plains headed for New York City, for Washington, DC, for Boston. In a wide swath running north from Texas through the Dakotas, an estimated seven thousand people died and more than two million left their homes. Some deaths came directly from the wind, from tornadoes that ripped across the land, eating barns and houses and cattle and people, but mainly the deaths came from the dust. The dust, inhaled, led to inflammation of the lungs. Dust pneumonia affected the very old and the very young and everyone in between. Healthy men and women experienced chest pains and shortness of breath and incessant coughing. Some died sooner than others. Some babies never left their cribs, while others carried the lung scars of dust pneumonia into old age, along with memories of poverty and uncertainty and bitter loss.

The Red Cross passed out dust masks. The masks, catching some of the dust, turned black within hours. Lungs, unseen, also turned black.

The dust found its way down throats, into stomachs and intestines, contributing to the discomfort of those affected, adding to the general malaise.

The dust did not discriminate between people and animals. Cattle and chickens and sheep died, lungs and guts full of dust.

Plants, sandblasted and buried, died.

The landscape itself died, its pretty creeks crusted with muddy grime, its fences trapping piles of powdered earth, its houses and its neatly plowed fields buried under moving dunes.

Neatly plowed fields stood at the heart of the problem. Prairie plants—the grasses and forbs that had grown there for millennia, grama and buffalo grass and galleta and their companions—once held the soil in place. Before the plow arrived, whether it rained or not, prairie plants persisted, roots married to the earth, foliage blanketing the ground, roots and foliage together preventing soil particles from vibrating and bouncing and leaping and flying.

But after the plow, the old plants were gone. The drought-resistant prairie community, the community designed by nature to survive drought and wind, was gone. Wild plants with wild toughness were replaced by the tame plants of farmers, plants that were like house pets in that they needed special care, in that they needed water on a regular basis, and in their susceptibility to the heat.

A soil scientist once told the farmers that their method of production was "suicidal production."

When there was rain, when it was not too hot, the crops thrived. Like the prairie plants before them, thriving crops prevented the soil from taking wing. But when the rain failed, the crops failed. When the drought came—locally pronounced, even today by some, "the drouth"—the crops failed. Soil particles moved aloft.

In a land without shade, where the landlocked knew what it was to be surrounded by horizons, weather could be seen from miles away. In one direction, the sky was clear and blue. In the other, the sky was gone, swallowed by menace, overwhelmed by darkness that, as it approached, resolved itself into a boiling black mass, turbulence made visible by virtue of flying soil. When the black mass hit, it shrouded the sun. Seething with fine powder and gravel-size grains and everything in between, it peppered roofs and windows. It invaded homes, sneaking through invisible cracks and crannies. It rendered the air, even the inside air, unsafe. Babies had to be held close, protected, somehow isolated from the very stuff they breathed. Families inside their homes, stunned, watched and listened and waited and inhaled fine particles of earth.

One of the worst dust storms hit on April 14, 1935. "A huge cloud of black top soil," according to a reporter writing for the *Leader Tribune* in Oklahoma, "swooped down upon Laverne in the manner of a heavy cloud flattening out upon the earth and spread absolute darkness the like of which has never been experienced by most Harper county folk."

By 1935, most of the people of Kansas, Oklahoma, and Texas were old hands at dust storms. By then they knew these storms as "dusters." They had grown accustomed to sweeping uninvited soil from their homes. Shoveling unwelcome dunes out of roadways using tools kept handy in Model Ts had become routine. But this particular storm, the April 14 storm, the Black Sunday storm, was especially bad, even to the hardened citizens of the prairie states. It was after this storm that people began to use the term "Dust Bowl."

*US Department of Agriculture photograph of a man hunched over in the wind of a Dust Bowl storm.*

From the *Liberal News* in Kansas: "Some people thought the end of the world was at hand when every trace of daylight was obliterated at 4:00 PM."

From the *Amarillo Daily News* in Texas: "The billowing black cloud struck Amarillo at 7:20 o'clock and visibility was zero for 12 minutes."

From the *Lubbock Evening Journal,* also in Texas: "Hundreds of Sunday motorists were caught when the dense black cloud bore down upon them at a rate of 60 miles an hour. Many motorists who attempted to drive through the cloud of stinging gravel and sand found that static electricity, generated by the dust particles, had disrupted the ignition systems of their engines."

The wind picked up decades of toil and hopes and dreams and sent them flying. Its noise drowned the sobs of grown men. The wind swept away not only soil but also lives and souls and sanity. Beaten families abandoned babies on church stoops. Men and women, mad with anguish brought on by wind-borne dust, were institutionalized.

From the Weather Bureau at Dodge City, Kansas: "The wind was travelling at a speed of sixty miles an hour; when it struck, visibility was reduced to zero for a period of twenty minutes, after which time visibility was limited to ten feet or less, lasting for forty-five minutes, then visibility increased to fifty feet or more at sporadic intervals and thereafter gradually increasing until normal by nightfall."

The Weather Bureau, while hardly inept, was a bureaucracy more focused on the statistics of weather than on the science behind reliable forecasts. The men of the bureau could measure the speed of the wind that had just passed, but they could not predict what would come next, what new horror lurked two or three days out.

People responded to the wind with clubs. They clubbed

jackrabbits, which were blamed for grazing on the scattered plants that struggled in the drifting dust and for eating what little grain found its way into silos. Up to six thousand rabbits could be killed in an afternoon's work.

At about the same time, the US Forest Service, Civilian Conservation Corps, and Works Progress Administration put impoverished farmers to work. The farmers traded clubs for shovels. They dug holes and planted trees.

Planting started in Oklahoma with a single Austrian pine in 1934. By 1942, more than two hundred million trees stretched in thirty thousand separate belts spanning a cumulative distance of almost nineteen thousand miles. Among the trees planted were cottonwood and Chinese elm, black locust and catalpa, green ash and honey locust, mulberry and red cedar and walnut. The trees tripped the wind. The trees were part of a strategy to conquer the wind, to stop it from stealing soil and crops and human vitality.

Other changes, changes felt in the world of meteorology, accompanied the rabbit killing and tree planting. While the dust flew, aviators, many trained for the war in Europe, also flew. On June 27, 1943, on a bet, pilot Colonel Joe Duckworth and navigator Lieutenant Ralph O'Hair intentionally flew their plane into a hurricane south of Galveston, Texas. The plane was a single-engine AT-6 Texan, a tail dragger with a glass canopy that could, when the plane was not flying in hurricanes, be slid open. Duckworth apparently had so much fun that he landed, took aboard another colleague, and flew into the hurricane again.

Commercial aviation was also well established by the time of the Dust Bowl. By 1933, United Airlines was flying coast to coast, a journey that required twenty hours. In 1938, pressurized cabins found their way into the growing fleet of commercial airlines. By 1945, competition compelled airlines to advertise low fares.

Fliers could provide data from altitude, and the growth of commercial aviation also drove the demand for better forecasts.

Slowly, science from Norway came not only to the Great Plains but to the entire nation. The fronts of the Bergen School began to show up on weather maps. The school's methods may have relied on graphics and subjective judgment, but they worked. They gave people an inkling of the weather to come. They did not lead to nonsensical results.

But meteorologists did not entirely abandon mathematics. Despite the usefulness and relative simplicity of the Bergen School's graphical methods, some meteorologists held on to their faith in numbers. The Welsh meteorologist David Brunt, writing in 1939, unapologetically salted his weather textbook with equations. "I take the view," he wrote, "that meteorology should aim at being a metric science wherever possible, and that no physical theory can be regarded as wholly satisfactory which cannot be expressed in mathematical form."

The winds of the 1930s reshaped a wide swath of land from Texas to Canada. But there is nothing unusual about wind shaping the land. It turns out that the winds of vortexes, the winds so richly explained by Bjerknes and his followers at the Bergen School, the violent winds of hurricanes and tornadoes along with the ordinary winds of daily life, the swirling patterns of moving air seen in so many satellite images of clouds, all regularly shape entire landscapes. The only thing unusual about the wind-driven reshaping of the wide swath of land from Texas to Canada was that people had removed the region's native plants and then suffered when drought dried the land and wind moved the earth.

In well-watered regions — in most of the regions where people live — the action of water blots out the action of wind. In well-watered regions, the obvious erosion is from running water and, farther north, from creeping glaciers. But in dry lands, there are ventifacts and yardangs. There are pans and lag deposits. There are moving dunes.

Ventifacts are rocks scarred by wind-driven sand and gravel. The cuts and grooves indicate the prevailing wind direction. Ancient ventifacts carry a record of changes in the prevailing wind. Ventifacts on Mars show the direction of the planet's surface winds.

Some ventifacts take on the appearance of a stone mushroom. Wind-lifted sand and gravel smash into the lower part of the mushroom, but the upper parts, above the fray, too high for the heaviest sand and gravel, remain relatively unmolested. The stem of the mushroom grows skinnier with every storm.

Yardangs are wind-sculpted rock hills, often taller than they are wide, having the appearance of rough-hewn spires, sometimes tilted as if the rock itself bends in the wind. Some have wind-cut windows through them, like the eye of a needle. The wind gouges out loose soil and attacks the softer rock, excavating the earth around the harder rock and making it appear as though the harder rock grew upward, when in fact the softer ground retreated.

The word "yardang" comes from southeastern Europe, imported to the Western world by the Swedish explorer Sven Anders Hedin, a contemporary and possibly an acquaintance of Vilhelm Bjerknes. He traveled extensively through the arid regions of Central Asia. "The ground we entered after leaving the L.M [sic] site showed at first clear signs of extreme wind erosion," he wrote. "Yardang trenches were scooped out to a depth of eight to twelve feet."

Yardangs may be beautiful, but they are not always convenient. "We must pitch camp, as soon as we found a suitable place with level ground for the tent," Hedin wrote. "But such spots were rare in country so broken up by yardang formation."

A large sandblasted stone feature could be both a ventifact and a yardang. The lungs of Dust Bowl survivors, if they were made of stone, would be ventifacts.

Pans—flat expanses, sometimes bowl-shaped—form when the wind lifts material away. In lag deposits, coarse pebbles reside atop finer sand.

A discussion of features shaped by wind could fill a textbook. The simplest, on close examination, become complex. A ventifact may be an einkanter, scarred only on the windward face; or a zweikanter, scarred on two faces; or a dreikanter, scarred on three faces. The wind itself, changing the shape of the rock and eating away the ground beneath the rock, may make the rock tumble, exposing a new face to the wind.

A dune—a simple windblown hill of sand—can be dissected into its topset beds, foreset bed, and bottomset beds, with its slip face sloped sharply downward at an angle of repose determined by the size and shape of its grains. Constant winds over areas where limited amounts of sand sit on hard ground form crescent-shaped barchan dunes. Add more sand, and the result is transverse dunes that look like lines of waves approaching a beach. Add sporadic vegetation to catch some of the sand, and parabolic dunes form. Additional winds, winds from different directions, create star dunes.

Lewis Fry Richardson, were he still alive, probably would support mathematical analyses and modeling of dunes and ventifacts and yardangs. If so, he would not be alone. He might work, for example, with the likes of David Cocks, who finished a doctoral dissertation on the topic in 2005. "We may therefore consider the flow in the constant flux layer only," wrote Cocks, "assuming that we know the depth $d$ of this layer. Then the logarithmic profile in the undisturbed case is given by (3.1), so that the velocity $Ud$ at the top of this layer is," and so on, for 177 pages.

155

Before starting *Rocinante's* engines, while we can still access the Internet, we check PassageWeather one more time. We do not discuss the source of the data, that the forecast relies on reports coming in from Voluntary Observing Ships and from automated buoys and from satellites, constantly correcting itself as the real weather unfolds. We do not talk of Vilhelm Bjerknes. Neither of us mentions Lewis Fry Richardson. Right now, we want neither history nor theory. All we want is a reliable forecast. All we want is a forecaster's final confirmation, a forecaster's permission to cast off and head, once again, for Mexico.

Here it is, laid out on our computer screen, the results of a mathematical forecast—the entire Gulf of Mexico color-coded for a quick view, to save the user from having to look too closely. The Gulf is white and light blue and dark blue. There is no green or yellow or red, nothing that Beaufort would call a gale or a near gale or even a strong breeze. Dozens of little lines decorate the screen, some with feathers on their tails indicating stronger winds, but none with the flags that warn of storms.

We scroll forward through time. For two days, the forecast is updated every three hours. For the third day, every six hours. After that, out to seven days, it is updated every twelve hours. We see our future in three- and six- and twelve-hour intervals, and all looks good. We will ride in the company of easy winds blowing five to fifteen knots from the east.

In three days, we should be speaking Spanish.

We cast off, bringing in the forward dock line first and then the stern dock line. Free, *Rocinante* inches forward. I let the breeze pull her bow to the left to avoid a skiff tied to the next dock. By inches, I avoid bumping a piling and losing our dinghy's outboard, currently mounted on *Rocinante's* rail.

The channel, its edges abruptly rising to within inches of the surface, its narrow deep water clogged with weekend boaters, presents a challenge. Some of the other boats appear to be commanded by Captains Magoo, men whose self-confidence compensates for their poor eyesight. We steam inland to reach a wider spot in the channel before turning about to point our bow toward the sea.

Within ten minutes of casting off, *Rocinante* comes abreast of the rock jetties leading to the Gulf. We wave at the elderly men and women in their folding chairs along the rocks. They wave back. Before we clear the jetties, we let the foresail fly, and it catches the wind. The sail, billowing and full, will pull us to Mexico. I shut down *Rocinante*'s diesel and am rewarded by the sounds of water splashing against the hull, the wind, and the double shriek of a passing gull.

Lewis Fry Richardson, armed with his pencil and paper and slide rule near the front lines of World War I, produced a preposterous forecast. The pressure changes he predicted simply do not happen in the real world. In trying to understand what went wrong, he landed on problems with the observations that populated his mathematical models. The glaring error that led to an impossible forecast, he believed, "could be traced to errors in the representation of initial winds." He was especially suspicious of the observations—or the lack of observations—for winds at altitude, in the cells high above the ground, measured, for the most part, by balloons. "These errors," he wrote, "appear to arise mainly from the irregular distribution of pilot balloon stations and from their too small numbers."

He was wrong about why he was wrong. The cause did not lie with the initial winds.

In 2006, meteorologist Peter Lynch reexamined Richardson's forecast. Richardson, in *Weather Prediction by Numerical Process*, left detailed records—painfully detailed records—of the stepwise procedure that he had developed and applied. Lynch followed the procedure step-by-step, but with the aid of a modern computer and with a modern understanding of the atmosphere. Lynch noted that Richardson's methods were ideally suited for modern computers, and that Richardson's forecast, though flawed, was "self-consistent." Richardson's error, Lynch found, came from his failure to properly account for minor fluctuations that in the real atmosphere cancel one another out.

"The core of the problem," wrote Lynch, "is that a delicate dynamic balance that prevails in the atmosphere was not reflected in the initial data used by Richardson." Ignoring that dynamic balance—the balance, in the parlance of academic meteorology, between the pressure gradient and the Coriolis terms—sent the forecast into a rat hole.

"The consequence of the imbalance," Lynch added, "was the contamination of the forecast by spurious noise." The imbalance, Lynch saw, "would engender massive gravity waves with wildly oscillating tendencies."

Waves at sea are in the family of waves known as gravity waves. In an ocean wave, wind disturbs the water's surface. The wave grows. Gravity eventually pulls the wave downward, but not before upward momentum lets it grow a little beyond a point of stability. And on the downward stroke, momentum pulls the wave a little beyond a point of stability before the wind can lift it back up.

The atmosphere, like the ocean, is full of gravity waves. They form where air of one density meets air of a different density. They can be seen at times as parallel lines of clouds that look like ripples. They can be seen less frequently in clouds that look like breaking surf.

To understand why Richardson's forecast failed, think of a very short video clip of a rising wave. Pretend that you have not seen waves all your life, and that somehow you do not know, intuitively, that the wave will swell into a peak and then slide into a trough. Think only of the swelling wave. In your mind, you see that wave growing beyond experience, beyond belief. It grows preposterously large. If it is an atmospheric wave, the pressure change beneath it would be likewise preposterously large.

Richardson's error came from his failure to eliminate meaningless fluctuations—fluctuations that would, in the real world, cancel one another's effects. The error came from his failure to smooth his data, to ignore things that in the real world are mere noise, balanced and counterbalanced, in practical terms nothing more than short-lived events eliminated by other short-lived events, each real, each important in the realm of physics, but taken together, in terms of forecasting, unimportant and uninteresting. The failure made his mathematical world go haywire.

As the demonstration of a method, Richardson's forecast was nothing less than remarkable. But as a forecast, it was junk.

Small changes would have fixed the problem. Richardson's forecast was wildly wrong, but his process was remarkably close to the mark. And to be fair to Richardson, his oversight might be excused given that so much of his thinking was done in a war zone, often within earshot of artillery.

Decades would pass before numerical forecasting gained ground. Even if Richardson's forecast had worked, in the days before electronic computers it would not have found its way out of academic circles. The mathematics, barely manageable for a single forecast for a small area, could not be scaled up. Richardson dreamed of a forecasting factory, of a theater set up to mimic the three-dimensional chessboard of his model and full of people working away on calculations for each cell, but he knew it was just a dream. He knew that his methods were for the time being entirely impractical.

As it turned out, the practical methods of the Bergen School established by Vilhelm Bjerknes, born of necessity, reliant on graphical methods rather than mathematics, would dominate the three-day forecast until the middle of the 1980s.

Just beyond the jetties, the wind dies. The foresail relaxes, hanging uselessly. The mainsail follows suit. The wind, with a mind of its own, abandons us until late in the afternoon. When the sun approaches the horizon, the promised wind returns, blowing at close to ten knots from the east. *Rocinante,* riding flat and comfortable over smooth seas on a broad reach, her main- and mizzen- and foresails hanging far out over the starboard side, makes five knots. We dance across waves turned orange by the setting sun. Astern, Florida has disappeared, submerged below the horizon. The Dry Tortugas lay off the port bow, also below the horizon, but within easy reach.

This is where another couple almost lost their boat in gusts reaching seventy-two knots. They were not world passage makers, but novices, like us, who had set out from Galveston Bay. They dropped anchor in the Tortugas after just four days of off-shore sailing. They received a report of a mild low-pressure system that would bring favorable winds from the southeast. But then the forecast was revised. The winds were now predicted to blow a steady forty knots. The barometer dropped seventeen millibars in twelve hours—nothing compared with what Richardson had mispredicted, but a substantial enough miscalculation in the real world to wreak havoc on yachts anchored in front of the Dry Tortugas. The couple's mizzenmast—the small mast that sits aft in a ketch—suddenly broke free from the deck. Before the wind eased, their boat was lying on its side, hard aground, badly wounded but alive. They refloated the boat and limped to the mainland. They undertook repairs and set off again. "We are doing our best to avoid all the major weather events of the 21st century," they wrote afterward, as they sailed on across the Pacific to New Zealand.

But all of that is ancient history, an event that occurred in 1993. Today's forecasts are more reliable. The couple had sailed on the advice of a late-twentieth-century forecast, while we sail on the advice of a twenty-first-century forecast.

*Chapter 7*

# THE COMPUTATION

We ride the trade winds. Frigate birds ride them, too, perched in thin air astride wings that stretch seven feet from tip to tip. They seem designed for flight, with air-filled bones that weigh, collectively, less than their feathers. They can and do stay aloft for days at a time, sleeping on the wing. Although they return to islands to nest and rest, their primary habitat is the air. Our frigate birds, the ones we see today, probably come from roosts in Cuba, but they could just as well have come from farther afield.

For the past two days, we enjoyed Goldilocks winds and smooth seas, a combination seldom reported by sailors. At night our wake sparkled with phosphorescent plankton, glowing greenish specks caught up in the turbulence of *Rocinante's* movement.

The water beneath the waves is alive. Plankton live throughout the water column. The neuston, distinct from the plankton, live at the top of the water column. Some of the neuston swim just below the surface. Some cling to the underside of the surface, like tiny marine bats. Some stand on top of the surface, perched on water tension. *Rocinante*, in passing, slices through neuston habitat like a knife, momentarily interrupting unseen lives. She mixes the uppermost surface with the water below, mercilessly blending the epineuston with the hyponeuston, temporarily desegregating neighborhoods in a long narrow swath across the Gulf of Mexico.

Although I have yet to see one, the creatures of the epineuston include the sea skaters, the genus *Halobates,* among the few insects to call the oceans home. They are small-bodied but long-legged. Most of them live near the coast, but five species are found at sea, far from shore, subject to the winds, tossed about in waves and carried by wind-driven currents.

It may be that I have not seen them because I have not looked closely enough. One survey reported more than a hundred thousand skaters on a single square mile of ocean. Involved in my own epineustonic existence, I have failed to notice theirs.

I have noticed scattered Portuguese man-of-wars. For the past two days, they, like *Rocinante* and her crew, have enjoyed perfect winds and smooth seas.

Jellyfish-like, the Portuguese man-of-war is not a jellyfish at all. A single man-of-war is not even a single animal. It is a colonial siphonophore, a relative of the jellyfish but not a jellyfish, not in the class Scyphozoa or even the subphylum Medusozoa. In its taxonomy and phylogeny — in its appearance and genetic history — the man-of-war is closer to fire coral than to jellyfish. But all of them — the man-of-war, the jellyfish, the fire coral — reside in the phylum Cnidaria, from the Greek word meaning "nettle," and all carry the stinging cells for which the phylum is known.

The man-of-war is a colony that looks and acts like an individual. Parts of the colony develop into trailing tentacles that can stretch more than a hundred feet. Above the tentacles, another part of the colony develops into a buoyant, translucent flattened balloon, a gas bladder bluish in color, known properly as its pneumatophore. If one were to sample gases from the pneumatophore, the sample would contain surprisingly high levels of carbon monoxide — not carbon dioxide from respiration, but carbon monoxide. Darwin himself would scratch his head about the evolutionary advantages of pumping carbon monoxide into a gas bladder. Why not carbon dioxide, the normal product of cellular respiration? But he would immediately recognize the flattened bladder for what it is. He would know a sail when he saw one, and he would recognize the adaptive advantage of growing one's own sail. He might see, too, that some of the creatures have sails swept to the right, and some have sails swept to the left, causing them to drift in different directions relative to the wind, a wondrous expression of diversity.

The man-of-war spends its days afloat, its sail-like bladder buoying the trailing tentacles, the entire colony riding the breeze, sailing.

These past two days, we—*Rocinante*, her crew, and uncounted man-of-wars—have sailed with the wind more or less behind us.

If we were to turn around, to head east, back toward Florida, *Rocinante* would struggle to make way. But compared with the man-of-war, *Rocinante* is a handy upwind boat. The man-of-war, even more so than the old square-rigged ships for which it was named, cannot sail upwind at all.

Each year man-of-wars pile up on beaches, driven there by the wind and stranded like wrecked sailboats. Ashore, the pretty blue sail-like bladders attract beachcombers. The long tentacles—stretched out in the surf, lying on the sand, or mixed with heaps of drying seagrass cast up by the surf—sting unsuspecting bare feet, causing pain in people who have never seen the grandeur of the man-of-war under sail, out where it belongs, out of sight of land.

Around midnight last night, the winds increased. Now, at dawn, six- and eight-foot seas chase *Rocinante,* coming from behind, sending her in brief bursts surfing down wave faces. The electronic autopilot struggles with swells on this downwind tack, trying to overcorrect when the boat slides down waves. In overcorrecting, the autopilot strains the steel pins that connect it to the rudder post. One pin works itself loose, and another breaks in half.

The autopilot, when it works, provides occasional relief for the crew. When it fails, the crew members take the helm. Either way, we progress toward Mexico. Before breakfast, we make soundings, with depths of four hundred feet. We should reach Mexico by noon.

Among sailors, frigate birds are sometimes said to come just before bad weather. More analytical sailors might say that they ride the leading edges of fronts. In contrast, Robert FitzRoy wrote, "When seabirds fly out early, and far to seaward, moderate wind and fair weather may be expected." To him, seabirds flying inland are harbingers of a storm. But he might have been thinking of seabirds other than frigates.

FitzRoy believed that many animals have some sense of the coming weather. "As many creatures besides birds," he wrote, "are affected by the approach of rain or wind, such indications should not be slighted by an observer who wishes to foresee weather."

FitzRoy was not the first to notice that animals seem to sense coming weather. From Horace, nineteen centuries before Fitz-Roy: "I will by prayer, from sunrise, arouse the croaking raven before the bird that foretells approaching rain revisits the standing pools." And from Pliny the Elder, eighteen centuries before FitzRoy: "When land-birds, especially crows, wash themselves, and are noisy near pools of water, it is a sign of rain."

Croaking frogs predict storms. Cows lie down before it rains, shielding a dry piece of land, avoiding the necessity of bedding down in a puddle. Watch the bees and butterflies, too. If sheep huddle together, expect a cold front. If ladybugs swarm, expect heat. And somehow groundhogs notice their shadows only in years of long winters.

In 1863, the year of FitzRoy's *Weather Book*, a Dutchman

published another book that included meteorological signs. Like FitzRoy, the author wrote of animal predictions, dwelling for a moment on the story of a French spy in Utrecht who predicted a severe winter based on the movements of a spider. The French were hoping to invade the Netherlands, but they were stymied by rivers and soft bogs, conditions that would slow horses and foot soldiers alike. Hearing the spider-based forecast from their spy, seeing an opportunity in the impending cold weather, the French crossed frozen rivers in 1795 and moved swiftly across the frost-hardened land.

The Dutch author believed in animal forecasts, but not as much as he believed in science. "Let it not be supposed," he wrote, "that we would deny that many animals should not be possessed of a peculiar sensibility which enables them to appreciate future changes of the state of the weather; but their aptitude for this peculiarity is rather limited, and our meteorological instruments inform us more rapidly and surely of any changes of the weather."

The wind picks up spray from the tops of the waves, occasionally throwing it across *Rocinante's* cockpit. In the morning sun, the spray shines from a spiderweb. The web stretches eight inches between a mizzenmast shroud and a lifeline, between a cable supporting our aft mast and a cable that keeps us from falling overboard.

Charles Darwin, during his journey under FitzRoy's command, saw spiders at sea. "In the evening all the ropes were coated and fringed with Gossamer web," he wrote. "I caught some of the Aeronaut spiders which must have come at least 60 miles."

He wrote of wind-borne spiders more than once. They are often called ballooning spiders, though they use silken strands of webs more as kites than as balloons. "On several occasions," he wrote, "when the *Beagle* has been within the mouth of the Plata, the rigging has been coated with the web of the Gossamer Spider. One day (November 1st, 1832) I paid particular attention to this subject. The weather had been fine and clear, and in the morning the air was full of patches of the flocculent web, as on an autumnal day in England. The ship was sixty miles distant from the land, in the direction of a steady though light breeze. Vast numbers of small spiders, about one-tenth of an inch in length, and of a dusky red colour, were attached to the webs. There must have been, I should suppose, some thousands on the ship. The little spider, when first coming in contact with the rigging, was always seated on a single thread."

I have no idea if the spinner of the web on *Rocinante*'s rigging was a stowaway picked up in Florida or a midcrossing boarder, a Cuban aeronaut. All I see now is its web. The spider itself is gone, maybe flown away for better hunting elsewhere, or maybe hiding, sensing bad weather. But I do know that more than one kind of spider flies on gossamer strands. The practice seems commonplace among small spiders. They have been captured at altitudes of sixteen thousand feet, they have traveled as far as one thousand miles, and they can sail through the air for weeks on end.

*Rocinante* shares the wind with more than one kind of sailor.

Biologists, forever classifying, tirelessly trying to categorize all things living and all places where they live, sometimes talk of biomes. The Great Plains form a biome of grass. The Arctic tundra is a treeless biome underlain by frozen ground. The tropical rain forest is a biome of remarkable diversity. In the early 1960s, biologist and Himalayan explorer Lawrence W. Swan described what he called the aeolian biome, named for Aeolus, the Greek god of the winds.

Swan looked at spiders in the shadows of Mount Everest above eighteen thousand feet. He watched them feed on flies and relatives of insects called collembolans, or springtails. "But how," he wondered, "do the flies and Collembola survive in an otherwise barren world of rocks and ice?" Under normal circumstances, young flies and collembolans relied on decomposing plant matter for food, but here, halfway to the stratosphere, there were no plants to decompose.

"The wind," he wrote, "brought nutrients from far away to feed established insect populations."

The normally close relationship between plants and animals could be maintained at a distance, he realized, facilitated by moving air. Wind could deposit nutrients on alpine snow, there to be processed by algae, the algae to be eaten by snow worms. And the wind, blowing across the ground, could stall at the edges of cracks and ledges, dropping its contents where they might be found by young flies and collembolans. The material carried on the wind could be dust, pollen, tiny insects, seeds, spores, or fragments of plants and animals. Or even the remains of seawater. "Small bubbles of the ocean surface," he wrote, "concentrate oceanic organic matter, and this concentrate is what is elevated into the air when bubbles burst and then is carried by the wind to add nutrients to the snow and the aeolian biome."

The stuff of the neuston rides the wind to high mountain slopes.

The aeolian biome supports more than just spiders and insects and collembolans. The lizard *Sceloporus microlepidotus* sometimes lives above fifteen thousand feet in Mexico, above the tree line. It is not a lonely existence. It lives on insects blown in from the tropics of Veracruz. It shares ground with snakes and, of all things, salamanders.

Elsewhere, on other high mountains, deer mice and shrews live on insects swept up by rising currents of air. On a volcano in

Hawaii, crickets survive on the products of the wind. On bare hardened lava in the Canary Islands, another group of spiders and insects rely on the providence of air in motion. Volcanoes in general, sterilized by explosions and flowing molten rock, repopulate thanks to the wind. Forty-three species of spiders parachuted onto Mount St. Helens during the first two summers of its latest stage of life.

It is often said that plants invaded the land, followed by animals. If said within earshot of Lawrence Swan, such a statement might have attracted rebuttal. Swan might have pointed out that the wind would have carried oceanic nutrients ashore long before plants crept up from the intertidal zone. Animal invaders from the sea did not need plant predecessors.

"I presume," Swan once wrote, "that the first invaders of the land were facultative aeolian heterotrophs from the seas." In plain English, the first animals to crawl out of the water did so to feed on detritus dropped by the wind.

Where others saw biological wastelands, Swan saw overlooked and underappreciated life. "The word aeolian," he wrote, "could also replace such mapmakers' terms as barren, glaciated, and snow or ice used to mark the lands beyond the tundra and that imply a misleading sterility."

An airborne connection between those who need nutrients and those who supply nutrients is not without risks. Sample the air, and more will come out than ballooning spiders and bits of blowing leaves and tiny insects and nutrients. Also riding the wind is the detritus of man, airborne rubbish, a combination of poisonous gases and noxious particles. It includes sulfur oxides and nitrogen oxides and lead and mercury. It includes fine particles less than three microns in diameter, the size of some bacteria. Some of the particles consist of intentionally deadly stuff, pesticides dreamed up by chemists for the purpose of killing those organisms considered, for one reason or another, undesirable.

As Swan put it: "The living things in these fragile aeolian ecosystems — life that holds only a tenuous grasp on existence from the air — may also be among the most endangered populations threatened by atmospheric fallout."

When *Rocinante* summits waves, the tops of hotels on Isla Mujeres pop up above the horizon, the first visible sign of Mexico. As *Rocinante* slides down the wave faces, the hotels disappear. We make six knots through the water, but only four knots of headway, held back by the Yucatán Current.

Soundings come and go at first, our sonar finding the bottom and then losing it again, confused by the seas. Soon enough, we show a steady depth reading. As we move forward, the bottom slopes upward, climbing toward three hundred feet. We have sailed onto the continental shelf. Less than an hour passes, and a mere one hundred feet of water separates the bottom of *Rocinante* from the bottom of the sea. During that time, we leave the Yucatán Current behind. We pick up almost two knots of headway.

The waves, still from behind, come at eight-second intervals, some bigger than others. When the boat begins her downhill slide, she wants to turn to the left. The wheel requires attention to keep her more or less on course and to keep her sails full. As we sail up the back of a wave, I bend my knees to accommodate the motion of the boat, one hand on a shroud for stability and one hand on the wheel, eyes moving from the wind vane at the top of the mast to the waves to the compass to the line of hotels growing on the horizon. At the crest, I turn the helm clockwise, anticipating the boat's tendency to swing to the left as she rides the wave.

With eight-second intervals between wave peaks, the knee-bending, shroud-grabbing, wheel-turning exercise recurs 450 times in one hour. During that same hour, the hotels grow entirely visible, as does the green vegetation behind the beaches and the white sand of the beaches themselves.

Just north of the island we cross a reef, charted at ten feet. The larger waves break as they cross the reef, their crests collapsing on themselves, becoming suddenly white.

Captain John Voss, the man who sailed a dugout canoe most of the way around the world, wrote of sailing on breaking waves in New Zealand, bragging about the experience, claiming he had shown something new to the New Zealand farmers at the end of the nineteenth century. I have no bragging rights. The breaking waves leave me nervous, jittery.

Clearing the reef and the tip of Isla Mujeres, we turn to the left and find ourselves in the lee of the island, abruptly protected from waves and wind. We lower the mainsail. We furl the jib until it is not much bigger than a handkerchief. With our engine running but not engaged, we sail with sand six feet under the keel and coral shoals to port and starboard. The channel takes us close enough to shore that we can see the faces of tourists on a pier, watching our maneuvers.

We pass a ferry dock and a navy dock. We roll in what is left of the jib and continue under power into the bay, protected from every direction but the northwest, dropping our anchor onto Mexican sand in ten feet of clear blue water. We share the anchorage with twelve other boats, ranging in size from twenty-eight feet to fifty feet and in condition from immaculate to barely afloat.

Ashore to the west, a narrow spit of mangroves growing in sand defines the edge of the bay. Beyond the mangroves, the hotels of Cancún crowd the horizon. But it is not the hotels that draw my eye. It is the wrecks that line the shore of the bay, awash in the shallows beneath the mangroves. Water rolls across two steel hulks, and farther up, touching the mangroves themselves, a sailboat lies on her side, someone's dream just beyond the edge of the anchorage, blown ashore, her rigging gone, her hull green with algae, clear blue ocean lapping at will in and out of her dark interior.

In the 1960s, before the Yucatán was targeted as a tourist destination, before Isla Mujeres was developed, the now sprawling city of Cancún reportedly supported only 3 residents, caretakers on a coconut plantation. Down the road, Puerto Juárez supported another 117 people. But by 1980, coordinated development was under way. A tourist zone rose from the coral reef that fringed the shore. El Centro, where the newly arrived residents lived, rose up with the tourist zone, following, at least in principle, a master plan of its own. By 1988, the population pushed past 100,000, almost all of them supported directly and indirectly by the tourist industry's 39 hotels and 10,000 rooms.

That year, Hurricane Gilbert came to town. Rear Admiral Beaufort would have ranked Gilbert's winds at level twelve, winds that "no canvas could withstand." But that description would apply to any hurricane. In Beaufort's mind, a hurricane was a hurricane. In contrast, the Saffir-Simpson scale—created by an engineer and a meteorologist to categorize hurricanes based on wind strength—subdivides them. A Category 1 hurricane, according to this scale, is "very dangerous and will produce some damage." By Category 3, "devastating damage will occur." At Categories 4 and 5, "catastrophic damage will occur." Gilbert came ashore in 1988 as a Category 5 storm.

Authorities issued the first warnings on September 13 at ten in the morning. Evacuation of the tourist zone began. By evening, five thousand tourists had been whisked away. The second warning came at three in the afternoon. By then, some of the residents, seeing the evacuation of tourists, were already gone.

During the evening, walking erect became a challenge. Driving a straight line became a challenge. That night, at eleven, a third warning was issued. People continued to flow out of the city. An estimated fifty thousand fled.

Some chose to stay. Later, many of them told interviewers that they had stayed to prevent looting. In the worst neighborhoods, the lack of confidence in law enforcement confounded the fear of the wind. Those with the least to lose feared the loss of property enough to risk all.

Neighborhoods at higher elevations fared better than those at lower elevations. High-quality houses, those built from *concreto* and *piedra* and *ladrillo*, from concrete and stone and brick, fared better than lower-quality houses, those built from *palmas* and *cartón* and *metal corrugado*, from palm trees and cardboard and corrugated metal.

The strongest winds hit after midnight and before dawn. The Mexican National Weather Service reported gusts of 218 miles per hour and sustained winds of 179 miles per hour.

Men died trying to save shrimp boats. A family perished when crushed by a collapsing wall. A baby drowned.

After daybreak, an unnamed army officer at Cancún's city hall commented to reporters on relief efforts. "We can't do it yet," he said. "The wind would blow them away."

Isla Mujeres protects the anchorage from the east wind, reducing it to a light breeze. We launch our dinghy and row ashore over thick beds of seagrass. We walk along the island's main road to a street-side restaurant. Tourists drive by in golf carts. A local man sells sugary frozen pops from a tricycle with a cooler. Schoolchildren, uniformed in blue and white, sit in the shade, waiting. Iguanas lie on concrete rubble in vacant lots.

The Mayans knew the region around Isla Mujeres as Ekab. They built a stone temple on the windswept southern tip of the island, where sea breezes blow through caves that penetrate the high bluffs, their walls wind-polished, making them ventifacts. At night, flames burning in the temple may have guided boats to shore. These temple fires may have guided Mayan traders, travelers, and fishermen working on a reef not yet damaged by the loving devotion of the tourists who would eventually come in throngs.

In 1517, Francisco Hernández de Córdoba sailed here from Cuba. In about 1566, Friar Diego de Landa recorded the still-living memories of Córdoba in his *Relación de las Cosas de Yucatán*. "During Lent of 1517 Francisco Hernández de Córdoba sailed from Cuba with three ships to procure slaves for the mines," Landa wrote, "as the population of Cuba was diminishing. Others say he sailed to discover new lands. Taking Alaminos as a pilot he landed on Isla de las Mujeres, to which he gave this name because of the idols he found there, of the goddesses of the country, *Aixchel, Ixchebeliax, Ixhunié, Ixhunieta,* vestured from the girdle down, and having the breasts covered after the manner of the Indians."

Córdoba found gold objects in the temple and took them. According to Landa, "The Indians marveled at seeing the Spaniards, touching their beards and persons." But the friar was a murderer and a book burner as well as a man of the cloth, and in later life, when he wrote his account, a guilty conscience might have guided his pen.

Although Isla Mujeres, the Island of Women, is less than a mile across at its widest, on this side of the island the air barely moves. Whatever wind comes from the east is entirely blocked by the island's narrow band of sand and trees and houses. The island is smaller than one of Richardson's cells, but it is big enough to alter the wind.

"This land," wrote Landa, "is very hot and the sun burns fiercely, although there are fresh breezes like those from the northeast and east, which are frequent, together with an evening breeze from the sea."

He also mentions the winter winds, the kinds of winds we tried to ride south from Texas. "The winter begins with St. Francis day," he wrote, "and lasts until the end of March. The north winds prevail and cause severe colds and catarrh from the insufficient clothing the people wear."

Later we walk toward the northwestern tip of the island, past the tiny naval base and the ferry dock, and we turn right, toward the sea. We walk past souvenir shops and *tequilerias,* small budget hotels, storefronts hawking rental scooters, and scattered homes stuck between the shops, their inhabitants happily going about their lives with open windows and wide double doors, ignoring passersby. The still air smells of food and motorcycle exhaust and rubbish.

In narrow alleys between buildings, breezes blow. As we get closer to the sea, the wind blowing through the alleyways grows in strength. Closer still, and the air moves everywhere, twisting and turning around obstructions. The smell blows away.

We see damaged buildings, reminders not of Gilbert but of its more recent cousin, Wilma. One hotel remains entirely abandoned, a mere shell.

The wind at the beach blows a solid twenty knots, picking up sand, piling it against painted concrete barriers that separate the beach from the street. Whitecaps cover the sea.

We walk along the beach, back toward *Rocinante,* turning inland only when we are close to our dinghy. The wind dies almost immediately. We smell something that could be sewage or a lagoon at low tide, or both.

Landa, in describing the Yucatán, wrote of tides that left seagrass exposed and of "much mud." He, too, may have breathed this same foul air.

In today's world, computer models, simple and complex, forecast air pollution plumes. Air pollutants disperse with the wind. Dispersion models forecast the plumes, showing a pollutant's footprint as it moves across the landscape.

Take, as an example, an especially nasty chemical, chlorine gas, often stored under pressure as a liquid in tanks and used to treat drinking water. Chlorine is a wonderful disinfectant, a low-cost killer of bacteria and viruses, a remover of iron and manganese and hydrogen sulfide. Is there a risk of brain-eating amoebas in your water? Not, it turns out, if you add the right amount of chlorine.

On the downside, chlorine in water may generate carcinogens. More to the point with regard to moving air, if pure chlorine escapes from storage tanks, it vaporizes, going from liquid to gas as the pressure drops. The gas, heavier than air, flows as a low-lying cloud, an evil yellowish-green fog smelling of bleach. In contact with water—sweat, for example, or the tears that bathe the eyes, or the moisture coating the insides of lungs— chlorine becomes hydrochloric acid. Hydrochloric acid in the eyes does not improve one's vision. Hydrochloric acid in the lungs does not improve one's stamina.

In sublethal exposures, expect blurred vision. Anticipate a burning sensation on the skin, accompanied by redness and blistering. Notice that burning sensation extending into the nose and throat. Prepare for a tightening of the chest and difficulty breathing. If possible, get out of the cloud. Get above it. Run upwind.

To understand the risk of stored chlorine, the mathematical models use methods similar to those that started with Bjerknes and Richardson. A computer generates a grid around a storage tank, plopping down half-mile-square cells extending outward for ten miles or more. The cells might be smaller if the terrain is complex. The computer, using average weather data, looks at air movement through the cells in short increments—say, thirty minutes at a time. In this manner, the computer identifies areas at risk.

Say, for example, that someone drives a forklift into a chlorine tank. These tanks are far from fragile. The driver has to hit the tank hard with the fork to rupture it. But say the fork pokes through. Say the hole is big enough to release 2,000 pounds of chlorine from the tank in one minute—that is, the hole is big enough to let loose about 160 gallons in one minute. As those gallons flow, they vaporize, turning to gas.

In an accident like this, modeling is not needed to understand that the driver will die. But modeling is needed to understand where the gas will move from there. If, say, the wind is barely blowing, a mere three knots, four miles per hour, and if, say, the terrain is level and open, without buildings or trees, the model might show chlorine traveling one mile in fifteen minutes. If the wind holds steady, the footprint will be narrow and cigar-shaped, but the model, being precautionary, will allow for a wind shift of, say, thirty degrees in either direction. The footprint of risk, the footprint of chlorine concentrations that could be immediately dangerous to life and health, will be V-shaped, a mile long and, at its outermost extent, almost a mile wide. That model output could be dropped on top of a map. The map might show schools in the plume, or day care centers, or nursing homes.

In 2005, a train derailed in Graniteville, South Carolina. More than a hundred thousand pounds of chlorine—something like eight thousand gallons—spilled out over the tracks. More than five thousand people were evacuated. More than five hundred people were treated at hospitals. Nine people died.

During World War I, chlorine gas was among the gases sent into the trenches. There was also the slower-acting phosgene, and there was mustard gas, ethyl bromoacetate, and methylbenzyl bromide.

Lewis Fry Richardson, working on his weather calculations near the front, was familiar with these gases. Like his comrades, he probably preferred winds blowing toward the enemy's lines. Compared with his comrades, he might have had a better feel for impending wind shifts.

Richardson believed in neither the morality nor the efficacy of war. His service in the Section Sanitaire Anglaise, as one of fifty-six men driving twenty-two ambulances, allowed him to support his country without bearing arms. Working on his calculations so close to the front lines, he must have understood the military importance of weather forecasting. This could not have been a welcome understanding for a Quaker pacifist.

Our row back to *Rocinante* is interrupted by a slender man in a small sailboat waving us over. He is a talkative fellow. His accent is vaguely French, but he is from Jordan. He bought his boat in Florida, for four thousand dollars. He holds strong opinions about the wind. Among his opinions is one that led him to intentionally sail into a squall line soon after he bought his four-thousand-dollar boat.

"It was a way, you see, of testing the boat," he says. The boat itself did fine. His girlfriend, aboard at the time, tumbled face-first into the boat's stove. She broke her nose. The memory of her broken nose brings a smile to his face. "It was a broken nose," he says. "Bloody, yes. Painful, yes. But serious? No."

He delivers all of this in minutes, an enthusiastic downloading to a stranger. While he talks, I row an occasional quiet half stroke to hold the dinghy in position.

He has been here in Isla Mujeres for a year, living on his small sailboat, trying to earn money to sail south to Panama.

He asks cheerfully if we have heard about the boat towed in today, demasted. We have not. He describes the boat, naming the make and model. It was towed in, he says, by the Mexican navy. When the boat's mast failed, it crashed onto plastic tanks of diesel fuel stored on deck. The tanks ruptured. Diesel fuel flowed into the boat's cockpit and somehow leaked into the boat's cabin. In a boat that never should have left sight of land, a boat built for weekends in bays and lakes, the crew wallowed in the waves, diesel fuel sloshing in the bilge and diesel fumes permeating the air.

A freighter, flagged down, pulled alongside. The Mexican navy was summoned.

The boat, he says, had sailed from Texas. And now the boat's owner is ashore, drinking. "He is more interested in beer," the man tells us, "than in finding a new mast."

He warns us of a wind from the northwest, forecast to be upon Isla Mujeres in a few days. "The bottom here feels like good holding ground," he says, "but it is only two feet of silt on top of limestone." When the wind blows, he claims, our anchor will skip along the bedrock, never burying itself deep enough to hold against the blow.

"All of these boats"—and here he makes a sweeping gesture—"will drag. The harbor will be a mess." His own boat hangs safely on one of the harbor's few permanent moorings. He suggests that we re-anchor, near the ferry dock. The dock will afford some protection from the northwest wind, but more important, the bottom there is thick sand, true holding ground.

We row away, downwind and toward *Rocinante,* and a moment later a plume of marijuana smoke catches up to our dinghy, pungent, not altogether unpleasant, covering the other smells of the harbor. The man in his boat, seeing us look his way, waves, and we row on.

Richardson, by World War II, was no longer forecasting weather, and his numerical methods were not in use. Between the wars, his book had attracted some attention. Science-minded meteo-rologists, including those of the Bergen School, recognized that the principles were sound, but also that, in practice, the calcula-tions were too difficult. Richardson's method, in a time before computers, could not keep up with the actual unfolding of events. Forecasts of the future could not be prepared before the future became the past. And there was the nagging reality of what Richardson called his "glaring error," the impossible results of his numerical forecast.

None of this drove Richardson out of meteorology.

Richardson left meteorology because the military annexed British weather forecasting. After World War I, the government had to decide whether it would maintain two meteorological services, one for civilian science and one for the military. Win-ston Churchill, appointed as the first secretary of state for air, weighed in on the side of a single service that would reside within the Air Ministry, which also controlled the Royal Air Force. Churchill, as he often did and often would, won the day. Richardson resigned, becoming a lecturer of physics and math-ematics, continuing his meteorological research on his own for a time. Eventually, though, he gave it up entirely.

*Lewis Fry Richardson abandoned meteorological studies over concerns that they were attracting attention from people interested in gas warfare. (Image from Wikimedia Commons)*

Later, his wife explained his decision. "Those most interested in his 'upper air' researches proved to be the 'poison gas' experts," she wrote. "Lewis stopped his meteorological researches, destroying such as had not been published. What that cost him none will ever know!"

Back on board *Rocinante*, I am at once elated, tired, and a little depressed—elated to have arrived here, tired after two weeks of irregular sleep, and depressed by the anxiety of incompetence.

In the cockpit, under the shade of our canvas bimini, I read from a 1978 copy of William Crawford's *Mariner's Weather*, a long-ago gift from my father. "When dissimilar air masses come together," it says, "the encounter is not a friendly merger."

I read about frontogenesis, the building of a front as two air masses meet. I read about occlusion, when a cold air mass overtakes and works its way under a warm air mass. I read about frontolysis, the failure of a front as two air masses move apart.

We call them fronts because Bjerknes worked under the shadow of World War I. We call them fronts as though they are battle lines on a flat battlefield. In fact, they are three-dimensional surfaces, changing constantly, bits of one air mass spinning off into the other, both air masses in motion, neither one seeking victory, yet both striving, sometimes violently, to restore equilibrium, on the largest scale striving to balance the heat of the tropics with the cold of the poles.

Lewis Fry Richardson, in a playful moment, wrote, "Big whirls have little whirls that feed on their velocity, and little whirls have lesser whirls and so on to viscosity."

Richardson gave up meteorology, but he did not give up his numbers. Forsaking weather because of war, he applied numbers to human behavior. For a time, he brought mathematics to psychology. "Psychology," he wrote, "will never be an exact science unless psychic intensities can be measured." He quantified the sense of touch using pinpricks. He quantified the longevity of mental images.

And then, in 1935, he wrote two pieces for the journal *Nature*. They presented mathematics—the kind of mathematics known in the field as "first-order coupled ordinary differential equations"—seeming to show that Germany, after World War I, would have to regrow its military. His equations predicted World War II.

He continued to throw mathematics at violence. He wrote *Generalized Foreign Politics* in 1939, *Arms and Insecurity* in 1949, and *Statistics of Deadly Quarrels* in 1950.

Thinking of the norther to come, thinking of the warning about strong winds and poor holding grounds, I stow my weather book and return to my shaded cockpit with another volume, another old book. A shadow of mold grows on the edges of its pages. Its title is *The Complete Book of Anchoring and Mooring*. Between its covers, it offers 359 pages of advice. There are many words and many drawings of anchors and anchor chains, and discussions of the angle between the seabed and the anchor chain and of the proper weight of anchors for different boats in different winds for bottoms of sand or mud or rock, and advice on swivels to prevent twisted chains. There are many right ways to anchor a boat, it seems, but even more wrong ways.

Heeding the advice of our Jordanian neighbor, we decide to relocate. My co-captain takes the helm while I raise the anchor. She maneuvers *Rocinante* across seagrass and over patches of sand to the old ferry dock. There, through the clear water, we see a sand and mud bottom, and the dock itself offers protection from the northwest. My co-captain circles twice, watching the depth sounder, and gives the command to drop the anchor. When it is on the bottom and tied off, she puts *Rocinante* in reverse, using the boat's power to tension the anchor line, digging in. We come to a stop, more secure in our new location, and my co-captain shuts down the engine.

During Lewis Fry Richardson's lifetime, weather forecasting, especially the forecasting of wind, would become integral to military operations. Wind impacted killing, and those who knew what the wind might do had an advantage over those who did not. Moving air affected military ships, but also observation balloons and biplanes and, later, entire air forces. It impacted artillery trajectories and the paths taken by bombs dropped from planes. It carried paratroopers past their landing zones. Its direction and speed determined how far a long-range bomber could travel before giving up any hope of returning to its base.

And one war after Richardson's war, there was D-day. Something like two hundred thousand men, eleven thousand aircraft, and almost seven thousand boats waited for what has since been called the most important weather forecast of all time.

In May 1944, the weather for an invasion was perfect, but the landing craft were not ready to go. General Dwight D. Eisenhower, in command, knew that he needed moonlight for airborne assaults, a low tide for landing craft, and not much wind. The first two he knew he could have on June fifth, sixth, and seventh. The wind was the unknown. Without good weather during his three-day window, he later wrote, "consequences would ensue that were almost terrifying to contemplate."

A Norwegian forecaster working for Eisenhower predicted deteriorating conditions. The upper air was becoming unstable. Using methods developed by the Norwegians, methods based on the principles of weather as they were then understood, methods grounded in physics and mathematics but dependent on graphical methods, he did not foresee invasion-friendly weather. But an American forecaster, also working for Eisenhower, though without the benefit of the advances set in motion by Vilhelm Bjerknes, disagreed. He believed that weather in the future was most reliably seen in weather in the past. Find an old weather map that resembles today's conditions and follow the progression of the past weather to understand how today's weather will progress. He saw nothing but good news. Based on fifty years of weather records, he believed that the Azores High, an area of high pressure that settles over the eastern Atlantic in the summer, would protect the English Channel.

Both the Norwegian and the American presented long-range forecasts extending out nearly a week into the future, notoriously unreliable in 1944. As the potential dates for invasion approached, the forecasters were expected to agree. Their forecasts were expected to converge on a common vision. But this convergence never happened. Instead, their disagreements intensified. Tension grew. The personalities of the forecasters clashed. The reserved Norwegian followed the ways of the Bergen School, combining theory with practicality. The brash American relied on methods that appeared practical but that in fact had little scientific basis. He believed in methods that Richardson, two decades earlier, had dismissed in the second paragraph of his book.

The American stuck to his belief in a calm fifth of June. The Norwegian disagreed. With data coming in from Canada, Greenland, Iceland, Ireland, and Britain itself, the Norwegian foresaw strong winds.

Later, one of Eisenhower's advisers remembered the experience. "If ever in the history of weather forecasting there was an occasion for unanimity of view and confidence in the outcome," he wrote, "this was it. Instead, here was a deep cleavage and uncertainty."

In the early morning of the fourth of June, with Eisenhower contemplating invasion on the fifth, the weather on the English Channel was calm. Briefing Eisenhower, the adviser expressed a lack of confidence in the forecasts. "Even for tomorrow the details are not clear," he told the general. "But we do know that the extension of the Azore anticyclone towards our southwest shores, which some of us thought might protect the channel from the worst effects of the Atlantic depressions, is now rapidly giving way."

The fifth of June was upon him. Eisenhower had to make a decision. He decided that the two hundred thousand men, eleven thousand aircraft, and almost seven thousand boats would hold tight.

The Norwegian's forecast came to pass. The wind howled.

But then the Norwegian, still using the methods of Bjerknes and the Bergen School, methods founded on theory and physics, saw an approaching lull. He saw a window of opportunity—not perfect, not long-lived, but enough.

On the other side of the English Channel, the Germans saw the strong winds but not the lull. German soldiers let down their guard. Field Marshal Erwin Rommel left his troops to present a gift—a pair of shoes from Paris—to his wife. Allied forces landed during the predicted break in the storm.

Aboard *Rocinante,* the predicted norther hits at dusk with thirty-knot gusts. She swings on her anchor. We watch the shoreline. We watch the anchor line. The shoreline stays more or less in place. The anchor line stays taut. Within an hour, gusts drop to twenty-five knots, and then twenty. We have a drink, snug on our anchor, secure and relaxed as we listen to the diminishing wind.

In 1922, Lewis Fry Richardson wrote of his dreamed-of theater packed with thousands of human computers, workers armed with pencils and slide rules in tiers of desks surrounding a central conductor, a foreman of sorts. Richardson knew it was just a fantasy. But by the end of World War II, others were not so sure. "The role of the enormous weather factory envisaged by Richardson with its thousands of computers," wrote meteorologist

Jule Charney in 1949, "will be taken over by a completely auto-matic electronic computing machine."

Charney was born in 1917. In 1948, barely into his thirties, he was leading the Meteorology Project at the Institute for Advanced Study in Princeton, New Jersey, with funding from the US Navy's Office of Research and Inventions. He and his colleagues at the institute were armed with ENIAC, one of the first real electronic computers, a machine that could be programmed for different purposes. Its 17,468 vacuum tubes, 7,200 diodes, 1,500 relays, and other components weighed 60,000 pounds. Five million hand-soldered connections let electrons flow through its innards. It drew more than 150 kilowatts of power, spawning rumors that the lights dimmed every time it ran. It was the size of a bus.

*US Army photograph of ENIAC, with programmers Betty Jean Jennings at left and Fran Bilas at right. (Image from Wikimedia Commons)*

What Richardson had done with a pencil and paper, Charney and his colleagues did, in principle, with ENIAC. They applied mathematics to the atmosphere to generate a forecast, and the mathematics they used required them to divide the atmosphere into a three-dimensional chessboard. They did not, however, use the same equations that Richardson had used. In less than three decades—the time that had passed since Richardson's calculations—the science of meteorology had leaped forward. Charney and his colleagues took a somewhat different route to numerical forecasting, using knowledge that had been unavailable to Richardson in 1922. Among the advances was recognition of something called Rossby waves, also known as planetary waves.

Rossby waves are but one kind of waves moving through the atmosphere. There are also, for example, sound waves. Sounds audible to humans travel as waves with lengths ranging from a fraction of an inch to more than fifty feet. There are waves with greater lengths, such as the waves that sometimes show up in clouds as regular ridges, white strips in a blue sky, each with a length measurable in miles. Rossby waves—planetary waves—are longer still.

An understanding of Rossby waves starts with an understanding of jet streams. These rivers of air reside at the top of the troposphere, the part of the atmosphere that contains weather, between about four and seven miles above the earth's surface. Jet streams were observed after Krakatoa erupted in 1883. They were observed again in 1933 by Wiley Post, the first man to fly alone around the world and a record setter for high-altitude flying. They were observed later still by World War II pilots who reported tailwinds blowing at speeds of a hundred miles per hour. In 1939, a German meteorologist named these winds *Strahlströmung,* translated as "jet flow" or, more commonly, "jet stream." Today, pilots plan flights around jet streams, avoiding the upstream slog but embracing the downstream race, moving not only horizontally around the flow but also vertically, purposely gaining or losing altitude to take maximum advantage of high-altitude rivers. When jet streams wander far astray, as they sometimes do, planes show up in normally empty skies.

Jet stream winds, just like other winds, are driven by air with a higher pressure flowing toward air with a lower pressure, but the direction of jet streams—high above the ground, well removed from the surfaces that cause friction—is driven by the spinning earth, by the Coriolis effect turning the air as it moves away from the equator. They form at the high-altitude juncture of air described in the 1700s by George Hadley—the meeting place of warm air from lower latitudes and cold air from higher latitudes. Hadley focused on a latitude of thirty degrees north—the latitude of Jacksonville, Florida; and Cairo, Egypt; and Shanghai, China. He did not realize it, but high above the land and the ocean waves, well beyond the region of interest in his day, a jet stream flows. The phenomenon repeats itself at higher latitudes, at sixty degrees, where it is called the polar jet.

In a simpler world, the polar jet would flow evenly, constrained in its course. In the real world, it bends and weaves, much like a river bending and weaving, with broad meanders that can stretch out for thousands of miles. These meanders, these waves, from one peak to the next or one trough to the next, might stretch across an entire continent. They became known as Rossby waves, after their discoverer, Carl-Gustaf Rossby.

Occasionally, the peak of a wave might break off, something like a river oxbow breaking away from the river itself, forming a long curved lake. But a broken piece of a Rossby wave, unlike an oxbow lake, keeps moving. A broken piece of a Rossby wave, escaping and headed to lower latitudes, carries weather with it.

In 1922, Richardson had no idea that waves as wide as a continent moved around the planet, more or less driving the weather

188

across the middle latitudes. Richardson did not know that Rossby waves, in terms of meaningful forecasting, dampened the effects of local conditions and small-scale changes. Richardson did not even know of the existence of Rossby waves. In his ignorance, he was in good company. Rossby did not explain the waves that took his name until 1939.

Along with ENIAC and their knowledge of Rossby waves, Charney and his colleagues were armed with Charney's own insights. In his 1948 paper "On the Scale of Atmospheric Motions," Charney complained of a prevailing view that the mathematics of weather must include atmospheric phenomena that have little impact on weather itself. He saw no reason, for example, to include sound waves in weather calculations. Typical gravity waves, too, might be ignored. In a paper necessarily full of mathematics, he explained that "the motion of large-scale atmospheric disturbances is governed by the laws of conservation of potential temperature and absolute potential vorticity, and by the conditions that the horizontal velocity be quasi-geostrophic and the pressure quasi-hydrostatic."

In these words—bewildering to the nonspecialist, deeply meaningful to the specialist—he framed what meteorologists know as quasi-geostrophic theory, or QG theory. In these words, he provided a simplifying framework that reduced several difficult equations to a single simpler equation. The mathematics of forecasting did not have to include every interaction. Just as forecasters with maps ignored some aspects of the atmosphere and emphasized others, Charney's mathematical approach ignored some aspects of weather—what he called "meteorological noise"—and emphasized others. The approach avoided the root cause of Richardson's error.

With the right mathematics, Charney and his colleagues might be able to see what forecasters saw in charts, a smoothed-out version of the weather. But with the right mathematics and an electronic computer, they could, maybe, see it better, and they could, maybe, propagate their vision further into the future.

And so they requested time on ENIAC, and when their request was granted, they put ENIAC to work on the mathematics of weather. Their hope, their desire, was a twenty-four-hour forecast.

They tinkered and fiddled. When ENIAC faltered, they replaced burned-out vacuum tubes. They handled stacks of punch cards — stiff paper cards speckled with holes that meant something to the computer. They played with different timing, running their model in one-hour and two-hour and three-hour chunks, seeing how coarse their approach could be before the output fell apart. Longer chunks meant fewer calculations. They settled on three hours.

The cells they used were 457 miles on each side. Big geographical cells meant fewer calculations. ENIAC liked simple. ENIAC, for all its bulk, was mentally deficient relative to today's machines.

A twenty-four-hour forecast required about twenty-four hours of work, including the handling of something like twenty-five thousand punch cards. At this particular time in history, at this particular time in the development of atmospheric science, the goal was not so much to calculate the weather before it actually occurred, but rather to demonstrate that calculating the weather was possible. In that regard, Charney and his colleagues were like Richardson had been when he was working away with his slide rule and pencil, eager only to show that mathematics could describe moving air. They, too, were following the original vision of Vilhelm Bjerknes, who, though he had turned away from purely numerical methods, saw mathematical modeling as the holy grail of atmospheric science. Bjerknes and Richardson and their followers, in the days before the first successful numerical calculations of the weather, in the days before Charney and his colleagues claimed success, wanted meteorology to follow the numbers. "If only the calculations shall agree with the facts," Bjerknes had written in 1914, "the scientific victory will be won."

Charney later said that the forecast had worked, at least on a large scale, with a few exceptions. But he also said that the mathematical model was flawed, that the input data—the information on the state of the weather at the beginning of the computations—contained errors, that the grid cells were too big, that there were challenges when it came to dealing with the edges of the model's cells. Charney and his colleagues knew that they had taken a first step but that more work was needed. In fact, a great deal more work was needed. The strangest thing about the ENIAC forecast may be that it actually worked, or at least it more or less worked.

Like the first aircraft, the ENIAC forecast hopped rather than flew. And also like that first aircraft, even though it did not work very well, it inspired more work. Numerical forecasting, suggested by Vilhelm Bjerknes in 1904 and tried by Lewis Fry Richardson in 1922, finally got off the ground in 1950. The ENIAC forecast, for all its faults, spoke volumes about weather forecasting. It showed the promise of the future.

In 1950, Richardson was almost seventy years old. Decades had passed since he had abandoned the weather, concerned about the role of forecasting in war. Jule Charney sent a copy of a paper describing the ENIAC forecast to him.

Richardson wrote back, calling the work "an enormous scientific advance," describing it as a giant step beyond the mistaken results of his own calculations. The ENIAC results seemed to show that accurate forecasts based on mathematical physics were within reach. All those years after leaving weather forecasting behind, his ideas had been resurrected and vindicated by a computer.

In his letter, Richardson did not indicate whether he knew

about ENIAC's history. Before calculating the weather, ENIAC had worked out artillery trajectories. It had also been used to assess the feasibility of hydrogen bombs.

Three years later, in 1953, Lewis Fry Richardson, still writing, still experimenting, still thinking about war, still an ardent pacifist, died in his sleep, not quite seventy-two years old.

*Chapter 8*

# CHAOS

Just after midnight, back at sea, my co-captain sleeps below. I have time on my hands to update my personal logbook. With a headlamp — red, to preserve my night vision — I scribble notes.

After some time in Isla Mujeres, we sailed southwest to Puerto Morelos, passing a wind turbine along the way, its blades spinning slowly, a giant pinwheel converting moving air into moving electrons just inshore from the beach. From Puerto Morelos we sailed southeast to Cozumel. South of Cozumel the shorefront resorts thinned out, and for many miles we sailed offshore from beaches backed by jungle, a Mayan coast, during daylight hours appearing not too different than it might have to Friar Diego de Landa in the sixteenth century. But at night, when the friar would have seen nothing but stars and the moon and occasional lights from fires, we saw the sky glow above Cozumel and Cancún and the scattered lights of villages and homes.

We anchored within the Sian Ka'an biosphere reserve near the village of Punta Allen, with its five hundred or so residents and its handful of tourists, most arriving by four-wheel drive over the thirty-mile-long dirt track connecting the coast to the highway. At Punta Allen, diesel generators provide electricity into the night, lighting the sky in a way that would have seemed most unnatural to Landa. But he would have recognized the wind, the reliable force of the trades.

From there we sailed south to Bahía del Espíritu Santo, the Bay of the Holy Spirit. Bahía del Espíritu Santo also has a village of sorts, but in two days anchored nearby, we saw no lights, no other boats, no commotion, just swaying palms, waves, and an occasional bird. The human residents must come seasonally, probably for the lobster fishery. Scattered concrete trays—lobster condominiums—hid beneath *Rocinante*'s keel in the sandy anchorage between the reef and the village. Each tray was three or four feet on a side and raised a few inches above the sand bottom. The lobsters hid beneath the trays, waving their antennae at the sea.

On the second evening at Bahía del Espíritu Santo, we watched a storm blow in from the east, thick cumulus puffs billowing upward, thickening, darkening. We cleared everything from the deck and started *Rocinante*'s engine, anticipating the tempest to come. A few drops of rain fell, followed by a fifteen-second deluge with winds gusting to twenty-five knots, and then both rain and gusts abruptly stopped, leaving only the trade winds and a blue evening sky. Later, the sun set over the Mayan jungle, turning red as it disappeared into the trees behind the beach.

Raising the anchor yesterday—day twenty-nine aboard *Rocinante*—our plan was to sail forty miles southeast from Bahía del Espíritu Santo and there to tuck into Banco Chinchorro, an offshore coral atoll—that is, a circle of coral, the remains of a sunken volcano or seamount, its innards a lagoon. We had planned a day sail followed by a night anchored within the peaceful lagoon of a tropical atoll, but the wind blew at twenty knots.

Darwin, sailing with FitzRoy aboard the *Beagle*, thought about coral atolls. He believed that they began as fringing reefs along the margin of an island and then the island sank, leaving the coral behind. "Now as the island sinks down," he wrote, "either a few feet at a time or quite insensibly, we may safely infer from what is known of the conditions favorable to the growth of coral, that the living masses, bathed by the surf on the margin of the reef, will soon regain the surface." Darwin was, in the matter of atolls, partly wrong. While islands sink, the sea level around them, at least since the Pleistocene, rises.

Banco Chinchorro is famous not only for its coral but also for its shipwrecks. Two galleons rest on the bottom there, as well as a ferry, thrown onto the reef in 2005 by Hurricane Wilma, a Category 5 storm with winds exceeding 150 knots. The Mayan coast is a hurricane coast.

Even at a mere twenty knots, at Beaufort's fresh breeze, we were not willing to test our luck in Banco Chinchorro's channels. We sailed on, resigned and not altogether unhappy with a night at sea, opting not to add one more wreck to the atoll's collection. Sailing is, among other things, an exercise in judgment.

Now, in the dark of night, the wind continues at a steady twenty knots. Forecast to blow fifteen, the trades have not dropped below twenty in the past fifteen hours. Closing in on one o'clock in the morning under a starlit sky, with my notes complete, I turn off my red headlamp. I stare into the gloom around the boat, feeling the invisible waves, judging their height at five feet. I look forward to Ambergris Caye. With luck, we will reach the island near dawn. With more luck, the wind and waves will be calmer there, calm enough to allow us to sneak through the narrow reef cut with its dogleg turn into protected waters. If not, our plans will change again, and we will carry on to the south, making the alternative entrance known as the Eastern Channel, adaptively managing our way down the Mayan coast.

Humans controlled fire at least two hundred thousand years ago and began farming about twelve thousand years ago. Five thousand years ago, someone crafted an Egyptian vase imprinted with the image of a boat under sail. Sometime before the twelfth century, someone in Persia rigged sails to a vertical spinning frame on land, creating a windmill, probably for pumping water or grinding grain. Early windmill blades spun on a vertical axis, like a merry-go-round with sails.

By the time windmills came to Europe, or at least by the time they were a common feature on the landscape, they were built on a horizontal axis, configured something like an airplane propeller. By 1390, tower mills, which may have first appeared along the Mediterranean, showed up in Holland, where the Dutch went to work on improvements. Soon after, the Dutch had windmills in place that would look at home on today's postcards. A cylindrical tower holds a four-bladed propeller frame, and the tower itself can be turned so that the propeller faces into the wind. Fabric sails can be rolled out onto the propeller frame or rolled in, as the wind allows.

The sails of these windmills were not simply catching the wind like a sock would catch the wind. The sails were cambered along the leading edge and twisted along their length. They created lift from the flowing air, like a sailboat with the wind abeam, skipping through the water far faster than a sailboat with the wind astern.

Inside, the Dutch installed wooden cogs and rings. The inside of a Dutch windmill in action, with the deafening noise of wood on wood and the complex motion of spinning wheels and belts, all transmitting power to a water pump or a grinding stone or the blades of a sawmill, would have pleased Rube Goldberg.

Flash forward a few hundred years, to 1888. In Cleveland, an inventor named Charles F. Brush erected a 6-story wrought iron tower behind his mansion. Near the top, he mounted 140 blades, each extending more than 25 feet out from a central axis. The blades were made of cedar. He was not interested in pumping water or grinding corn or sawing logs. Brush wanted electricity. Energy from his spinning windmill spun the shaft of a dynamo. The dynamo was connected to 408 batteries in Brush's basement. The batteries were connected to 350 lightbulbs and 2 electric motors.

Brush's windmill was the first to generate meaningful quantities of electricity. No doubt this feat impressed Brush's wife and neighbors almost as much as the noise from the spinning cedar blades.

*The Brush windmill, built in 1888 in Cleveland, was the first wind-driven turbine in the United States to produce meaningful amounts of electricity. (Copy of an original image in the Charles F. Brush, Sr., Papers, Kelvin Smith Library, Special Collections, Case Western Reserve University)*

The wind now, at three o'clock in the morning, blows at twenty-five knots from the southeast. We sail with a reefed mainsail and jib, making a steady five knots through the darkness. To our right, beyond the horizon, lies the coast, a lee shore in this wind. And I realize that we have now crossed the border. We have entered the territorial waters of the sovereign but disheveled nation of Belize.

A lee shore, to a sailor, does not provide comfort. The wind blows toward a lee shore, and a boat that has lost its ability to maneuver will be blown into a lee shore, there to be beaten to death in the shallows. From a short chapter called "The Lee Shore" in Melville's *Moby-Dick*: "But in that gale, the port, the land, is that ship's direst jeopardy; she must fly all hospitality; one touch of land, though it but graze the keel, would make her shudder through and through."

On our navigation screen, I see that the wind has pushed us toward the lee shore. Although our bow points more or less south, for a course that follows the coast, the wind pushes us not only forward but sideways. I adjust course, turning the bow a touch toward the east. I tighten the sails. *Rocinante* claws away from shore, but she is not her nimble self. Her jib is reefed to reduce sail area, rolled around its own forward edge. Where it bites into the wind, the rolled sail's edge creates turbulence. More turbulence tumbles off the bottom edge of the sail, filling the gap between the reefed sail and the deck. Turbulent flow does little to push a boat forward. Her mainsail, too, is reefed. Unreefed, the main and jib work well together. Reefed, they no longer work as a team, but as two separate sails.

The boat's efficiency diminished, we cannot point upwind as tightly as I would like. We come no closer than forty degrees to the wind. For a moment, I enjoy visualizing the air moving across the reefed sails, in my mind's eye seeing the invisible turbulence tumbling around the sails. But the enjoyment soon wears thin. We want to fly from the hospitality of the lee shore, but instead we claw along. In the darkness I learn one more lesson about *Rocinante*.

In this wind, the seas will be rolling through the narrow dog-leg channel that cuts through the barrier reef to Ambergris Caye, making the entrance dangerous, an unacceptable risk. We will have to sail on to the Eastern Channel, another thirty-five miles down the coast. But before long, we should be behind the Turneffe Islands, and they may offer some protection from the waves.

A fireball crosses the sky, not quite as bright as the one I saw over Cuba, but dramatic nevertheless, its greenish trail blinding me for a moment but leaving me delighted to be awake in the darkness of early morning, able to witness the vaporization of debris falling from space.

In 2010, at a meeting of the American Wind Energy Association in Dallas, George W. Bush received a standing ovation. "There's a big difference between talkers and doers," he told the crowd, "and here in the state of Texas, we are doers."

In this context, Bush meant doers of wind energy. Texas, following legislation put in place while he reigned as the state's governor, had become a leader in wind energy. The legislation could be remembered as a cynical polishing of environmental credentials by an ambitious governor with an eye on the White House. But it could also be remembered as a commonsense desire to take advantage of a resource with a long history in the great state of Texas.

Early settlers passing through the region would have heard of wind wagons, sailing ships on wheels. Farther north, in 1853 a man named William Thomas rigged sails to a wagon, eventually establishing the short-lived Prairie Clipper Company—short-lived because Thomas dramatically wrecked one of his wagons in front of a crowd of spectators.

*A wind wagon, also called a sail wagon, photographed around 1910. (Image from the Library of Congress)*

In 1860, another man, also farther north, rode a wind wagon five hundred miles before he met a whirlwind—a small tornado, a dust devil—that lifted his wind wagon off the ground, then dropped it, breaking the axle.

In Texas itself, in 1910, another innovator, H. M. Fletcher of Plainview, skeptical of conventional sails on wheels, erected a windmill on his wagon. With the assistance of gears and spinning shafts, he transferred the power of his windmill to the wheels of his wagon. The wagon journeyed from Plainview toward Amarillo, covering almost fifty miles before stalling on a hill.

By the time Fletcher's wagon stalled, ordinary windmills—not the kind harnessed to wagons—were a common sight on the Texas plains. Most drew water from one location and pumped it to another. Farmers used them. Ranchers used them. The railroads used them to fill the tanks that resupplied passing steam engines. Occasionally, they were rigged to oil wells. Wind pumped oil to the surface. And after Brush's innovation in 1887, they occasionally pumped electrons, bringing a few watts into homes otherwise entirely free of electricity. The Texas community of Midland, where windmills competed with homes for space, became known as Windmill Town.

In 1935, at a time when dust blew and settled into dunes at the bases of windmill towers, at a time when far too many willing workers sat jobless, Franklin D. Roosevelt launched the Rural Electrification Administration. The program strung wires to remote farms. The program centralized power production. The growing utility companies refused to hook up farms that still had operating windmills and forced the first generation of Texas wind turbines into extinction.

Windmills sat idle. They rusted. They were dismantled and sold as scrap.

Then, in 1981, Michael Osborne emerged as a wind enthusiast. With a background promoting music, followed by a stint as a solar power entrepreneur, Osborne found his way to Pampa, high in the Texas Panhandle. There, on a cousin's ranch, he built what is often called the first wind farm in Texas. For 1981, it was an ambitious project. With Texas bravado, he later called it the "second-largest wind farm in the known universe."

The wind power renaissance was not without problems. Wind turbines, in pumping electrons, also pumped noise. Neighbors found them ugly. Spinning blades killed birds and bats. Sometimes the turbines pumped more electrons than the local power grid could carry, and sometimes they pumped none at all.

In the absence of wind, the idle turbine blades cast long, stationary shadows. In the presence of wind, those motionless shadows sometimes remained in place, reminding their owners that moving parts need maintenance, that generating electricity puts tremendous stress on engineered devices, that things break.

Sometimes lightning struck. Sometimes the towers fell or the rotor blades flew apart and tumbled across the landscape.

Turbines with brake problems and leaking hydraulic fluid, with brake disks red-hot from friction, can and have caught fire, becoming giant flaming pinwheels.

Sometimes turbines just did not make enough money to pay their way. Osborne's Pampa project earned just under three cents per kilowatt-hour. "I really needed to be making around a nickel," Osborne said. He needed less expensive windmills or more efficient turbines or higher billing rates for electricity, or all three.

Osborne shut down his Pampa project, but there was no shutting down the enthusiasm for wind. Along came conventional oil companies and investment firms, outfits like Shell and BP and JPMorgan Chase and Goldman Sachs, funding and sometimes running their own farms. The farms grew larger and took on solid Texas names, such as Wildorado, Desert Sky, Capricorn Ridge, Sweetwater, and Buffalo Gap. Up sprang something like eight thousand turbines, making Texas the national leader in wind energy.

Throw into the mix billionaire corporate raider and hedge fund manager T. Boone Pickens, author of *The First Billion Is the Hardest* and owner of a ranch close to Pampa, Texas. Pickens, for a time, was a wind enthusiast's wind enthusiast. He loved wind not because he wanted to save the environment, but because he wanted to make money. "Don't get the idea that I've turned green," he told a newspaper reporter. "My business is making money, and I think this is going to make a lot of money."

Like Osborne before him, Pickens saw wind potential in Pampa. He promoted the town as the wind capital of the world.

He planned to produce four thousand megawatts of electricity with a little help from Pampa's wind. He dreamed of turbines on ranches in five counties generating power for more than a million homes. Local landowners would earn something like sixty-five million dollars a year in exchange for the use of their land, for the use of their wind.

A woman inquired about turbine noise. "If you're getting royalties from it," Pickens responded, "it might have a real pleasant sound."

But reality struck. The turbines never made a sound, pleasant or otherwise.

"I've lost my ass in the business," Pickens told a television audience in 2012. Wind itself is free for the taking, but the actual taking can be prohibitively expensive.

Understandably, Pickens, among the world's most successful investors, was disappointed, but he should not be ashamed. With his enthusiasm for wind, he was in good company.

"There is the power of the Wind, constantly exerted over the globe," wrote Henry David Thoreau in the November 1843 issue of the *United States Magazine and Democratic Review*. "Here is an almost incalculable power at our disposal, yet how trifling the use we make of it!"

And from Abraham Lincoln, in a lecture delivered in 1860: "I should think the wind contains the largest amount of motive power—that is, power to move things. Take any given space of the earth's surface—for instance, Illinois; and all the power exerted by all the men, and beasts, and running-water, and steam, over and upon it, shall not equal the one hundredth part of what is exerted by the blowing of the wind over and upon the same space." And this: "The wind is an untamed, and unharnessed force; and quite possibly one of the greatest discoveries hereafter to be made, will be the taming, and harnessing of it."

The wind farms of California, South Dakota, Oregon, Texas, New York, Indiana, and Pennsylvania would leave Thoreau and Lincoln impressed. Both men would be pleased and entertained to see wind farms rising out of the North Sea, the Irish Sea, and the Baltic Sea. They might smile at the notion of the Donghai Bridge wind farm off the coast of Shanghai. But they would be distressed by the number of turbines that do not spin and the number of farms standing stagnant, their backers out of money. Lincoln in particular would likely have an opinion about the labyrinth of regulations governing the survival of wind power. And, a lawyer himself, he might not be surprised by the presence of law firms specializing in wind energy.

Just before dawn, with stars in the east fading against a brightening sky, I hope for the wind to diminish, but if anything it increases slightly, gusting more often above thirty knots. The mathematics are simple: Double the wind speed, and the pressure on sails and rigging increases four times. Triple the wind speed, and the pressure on sails and rigging increases nine times. And so on. An experienced sailor considering the possibility of reefing sails, of reducing the amount of sail area exposed to the wind, spends very little time mulling over the issue. When the possibility of reefing occurs to an experienced sailor, he or she simply reefs.

We are not experienced, but we are already reefed.

The word "reef" comes from the old Dutch word *riffe* and the Old Norse word *rif*, both meaning "rib." A coral reef is a rib of coral at the bottom of the sea. A reef in a sail is a rib of the sail, a rib of bunched cloth, bundled together along a boom or wrapped around a stay, temporarily tucked away, out of the path of the wind.

I spin up the radar to scan for ships. Radar, now a standard word, started life as an acronym, adopted by the US Navy in 1940, for RAdio Detection And Ranging. "Spinning up" is not my terminology for turning on a radar set. It is the standard terminology of radar. It is the label on *Rocinante*'s radar panel. The mariner's lexicon did not stop growing with the demise of the square-rigged sailing ship. As sails gave way to steam and internal combustion, the men who had plied the seas grew old and died, but their tendency to invent new words persisted.

Spin up, and the radar antenna, inside its shroud, sends and receives radio waves. The outgoing waves bounce off objects, and the incoming waves become data on a screen. The data, in this case, show lots of backscatter close in. The backscatter is six- and eight-foot seas.

The principles of radar were demonstrated for the first time in 1904, the same year Bjerknes published his landmark paper on the promise of numerical forecasting. While Bjerknes was busy writing, a German inventor visiting Holland was invited to a conference on shipping safety. The inventor boarded a vessel called the *Columbus* to demonstrate his new toy.

"The trial on board of the *Columbus*," according to the notes from the conference, "though on very limited scale and with an unfinished apparatus, proved that the principle of the inventor is correct. Every time when, even at certain distance, a vessel passed, the apparatus operated immediately."

The invention could not measure distances—it could not give ranges, so it was not a true radar. But its potential was immediately recognized. At least one newspaper report remarked on a use beyond maritime safety, a use that might be thought of as the antithesis of safety, a use that would have troubled the still young Lewis Fry Richardson: "Because, above and under water metal objects reflect waves, this invention might have significance for future warfare."

By the time Richardson wrote the article in *Nature* predicting

World War II, most of the countries that would be swept into the conflict were experimenting with radar. By the end of the war, working systems were in place. It became possible to shoot at targets that were, but for radar, invisible.

During World War II, radar operators saw splotches on their screens, ghost echoes too big to be aircraft formations. They were rain clouds.

On radar, different weather systems have different appearances. Tornadoes, for example, look something like the number 6, with an ominously hooked tail.

In 1956, the US Weather Service acquired a Doppler radar system from the US Navy. Doppler radar looks not only at the returning pulses but at changes in the wavelengths of those pulses. If a target—a storm, in this case—is approaching, its movement squeezes the returning waves, shortening them, increasing their frequency. If a target is moving away, its movement lengthens the returning waves. Just as the sound coming from a moving train changes as the train approaches and then passes, the radio waves coming from the moving air of a storm say something about its motion.

The Weather Service started experimenting with the capabilities of Doppler radar. On June 10, 1958, in Kansas, researchers aimed their experimental unit at a funnel cloud. Inside that cloud, they detected rain moving sideways at just over two hundred miles per hour. Thirty minutes later, the tornado touched down in a small town called El Dorado. The very same winds that had carried the rain sideways at more than two hundred miles an hour destroyed 150 buildings and killed 15 people.

By 1971, thirty-seven military surplus radar systems used by the National Weather Service were joined by forty-eight purpose-built weather radar systems. None of these were Doppler systems. In the 1970s, Doppler radar remained experimental.

In the late 1980s, the government put contracts in place to develop Doppler radar for routine use. More than four hundred million dollars later, in 1993, Doppler radar was commonplace. Professional meteorologists appeared ecstatic. "It's incredible," said one. "We see things we always knew were there but couldn't see."

Overnight, television forecasters picked up Doppler radar images. What meteorologists could see, television viewers could see.

Not much more than a hundred years earlier, Vice Admiral FitzRoy had to defend his approach to weather forecasting against attacks by those advocating what amounted to an astrological approach, and now radar images supplemented numerical forecasts run on computers.

I adjust the range of *Rocinante*'s radar, zooming out. Close by, I detect only backscatter, but three miles out I detect a steady target. The vessel is not visible on my plotter's Automatic Identification System, its AIS. I see neither the vessel's name nor its destination. I track it for a few minutes on my radar screen and see that it is moving north. It will pass us to the west, between us and land. I turn off the radar screen, let my eyes adjust to the darkness, and spot a white running light. I wonder why it would travel so close to a lee shore in this twenty-five-knot wind. I cannot tell if it is under power or under sail, but either way, a lee shore is not a friend.

Sails were unfurled in ancient Egypt before the pyramids were built. They appeared around the same time in Southeast Asia. By the birth of Christ, sails were a common sight. By the beginning of the nineteenth century, eight thousand vessels could be seen in the port of London at any one time, a forest of masts bobbing near the city. London was but one of many ports sporting a floating forest.

Among the types of sailing ships, there have been barques and barquentines, caravels and carracks, junks and luggers, schooners and sloops and a dozen others, along with hybrids of every description. Of the types of sails, there have been lateen rigs and Marconi rigs, square rigs and gaff rigs. There have been canoes with a single sail and ships with two dozen sails, each with its own name and its own rigging.

The age of sail was largely, but not entirely, an age of men at sea. But occasionally women were smuggled aboard by officers and sailors. Others came aboard on their own and then dressed and acted like men, working alongside their unsuspecting male shipmates for months and even years. And there was Anne Bonny, Sadie "the Goat" Farrell, Jacquotte Delahaye, and Mary Read, all pirates. There were the wives of captains, too, accompanying their husbands on whalers and merchant vessels, at times raising families afloat. They faced hardships and met challenges. They showed resourcefulness and leadership. One, a Mrs. Jensen, whose husband's heart failed while their four-masted ship sat becalmed, was left to deal with a drunken crew far out at sea. Another, a Mrs. Follansbee, was becalmed along with her husband and their ship's crew in the Sunda Strait on April 22, 1838. A pirate vessel, propelled by oarsmen, approached. "Our cannon, swivel guns and pistols were soon got in readiness," she later wrote. She herself had been practicing for just such an event, but in this case she did not put her practice to the test. With the pirates less than a mile off, "a good breeze sprang up," she wrote, "and we were soon out of their reach."

A steam engine propelled a boat as early as 1783, when the *Pyroscaphe* steamed up a French river for fifteen minutes before breaking down. By 1813, a steam engine took a modified lugger named *Experiment* from Leeds to Yarmouth, an ocean passage. By 1880, a young Arthur Conan Doyle aboard a sail-powered whaler, prior to his invention of Sherlock Holmes, wrote of steam-powered ships between Greenland and Spitsbergen, wondering if the noise from their engines disturbed the whales. In 1895, Joshua Slocum, once the commander of trading ships propelled by sail but unfamiliar with and uninterested in steamships, found himself unemployed and unemployable. With nothing but time on his hands, he restored an abandoned oyster boat and sailed it, alone, around the globe. Other seamen, unemployed like Slocum, asked him, "Will she pay?" He made her pay via royalties from his book *Sailing Alone Around the World*, a bestseller, but even in 1895 he could not compete with steam-powered vessels in the transportation of goods or people. Slocum saw that the age of wind-powered shipping was all but over.

But to say that the age of wind-powered shipping was all but over is not to say that the age of wind-powered shipping is entirely gone. There is, after all, Jacques Cousteau's *Alcyone*, an expedition research vessel, 103 feet long, named for the mythological daughter of the winds.

*Alcyone* is not, to most eyes, a beautiful ship. Cousteau sacrificed beauty for functionality. Instead of sails, *Alcyone* sports two stacks. The stacks supplement her diesel power. They look like fat, stubby smokestacks, but they emit no smoke. Instead, they take advantage of a clever arrangement that uses a fan to suck air into them, creating an area of low pressure on one side, much as a conventional sail creates an area of low pressure by virtue of its shape. Devotees have claimed that the stacks—the turbosails—offer more thrust than conventional sails.

The *Alcyone* did not revolutionize sailing. Its turbosails have not caught on. And the *Alcyone* was not the first vessel to use what look like smokestacks instead of sails. The *Alcyone*'s turbosails were preceded by the Flettner rotor, named for Anton Flettner. Flettner's ship was the *Buckau,* which crossed the North Sea in 1925 and the Atlantic in 1926. His rotor stacks, though they looked something like the *Alcyone's* turbosails, worked on a different principle. Instead of sucking air up into the stacks, the stacks spun, taking advantage of the Magnus effect, the same effect that causes curveballs to curve. In a breeze, the spinning surface—whether of a stack or a baseball—carries a thin layer of air with it in the direction of spin. Because the spinning surface is moving forward, the spinning air on the side that spins away from the wind experiences slower wind speeds than that on the side that spins into the wind. Air piles up on one side and rushes away on the other, so high pressure develops on one side and low pressure on the other. The spinning stack or the spinning ball is pushed from the high-pressure side to the low-pressure side.

The Buckau, *powered by the wind via Flettner rotors (the tall stacks). (Image from the Library of Congress)*

From a largely forgotten professor named F. O. Willhofft, writing about Flettner's ship in 1925: "All that one can predict with certainty, basing the estimate with actual results obtained on the *Buckau* and on meteorological statistics, is that a motor-ship equipped with rotors will save not less than 25 percent on fuel, on the average, year after year, for the average trade route. I consider the Flettner rotor ship as a link only in the chain of evolution of the harnessing of wind power."

The Magnus effect is so strong that it has been applied to special airplanes sporting spinning cylinders where their more familiar brethren have wings. One such experimental plane was described in the November 1930 issue of *Popular Science Monthly*. "New mystery airplane," the headline reported, "built by three American inventors on barge in Long Island Sound, is tested in secret and is said to have made short flights." Where it should have had wings, it had "spools of metal, two feet thick." Based on the absence of such aircraft in today's skies, the secret tests of the mystery airplane must have come to naught.

Flettner's ship is gone, but a modern ship sails in its wake. From the German company Enercon on July 29, 2013: "The rotor sails on the ENERCON-developed *E-Ship 1* allow operational fuel savings of up to 25% compared to same-sized conventional freight vessels."

Enercon is not a shipping company. It is a German company that has installed more than twenty-two thousand wind turbines in more than thirty countries. Enercon built the Flettner rotor–equipped *E-Ship 1* to deliver wind turbines and, perhaps, to make a point.

Beluga Shipping, like Enercon, was a German company. Also like Enercon, it was interested in wind. Unlike Enercon, its primary business was shipping, not wind turbines. And unlike Enercon, it did not rely on Flettner rotors.

In 2007, Beluga Shipping launched its *Beluga SkySails*, a cargo ship more than four hundred feet long and fifty feet wide. Like other modern cargo ships, she rode the waves with the grace of a floating shoe box. In twenty feet of water, she would be hard aground. Despite her name, she carried neither sails nor Flettner rotors. But she could, in the right conditions, launch a kite. Her kite was twice the size of a baseball diamond. It was controlled through a computer console on the ship's bridge.

An executive at Beluga Shipping believed that the kite would save fuel. He believed that the kite would reduce emissions. And he understood the value of wind. "If you learned to sail at the age of eight, like me, sailing enters your bones," he once said, "and you get a feel for the wind, its power."

In 2011, through no fault of the kite occasionally flown in front of the *Beluga SkySails*, Beluga Shipping was declared insolvent.

Despite interesting designs, despite the promise of free energy from the wind, neither Flettner rotors nor turbosails nor Sky-Sails have caught on.

At daybreak, Ambergris Caye lies in our wake. We sail on.

Ambergris, the stuff from which the island took its name, was and is a valuable commodity, sought after by those who manufacture expensive perfumes. "Now this ambergris is a very curious substance," wrote Herman Melville in *Moby-Dick,* "and so important as an article of commerce, that in 1791 a certain Nantucket-born Captain Coffin was examined at the bar of the English House of Commons on that subject."

Found on a beach, ambergris looks something like a stone. It comes in different colors. It feels like wax but has an unusual smell, often described as "earthy." It sells for more than ten thousand dollars a pound. And it comes from the intestines of sperm whales.

Ambergris floats, and I watch for it when at the helm in daylight. I find none, and I have never met anyone who has improved on that record, but the small chance of finding does not, in this case, impact the pleasure of looking.

I see high clouds to the east, cirrus, broken into lines like breaking waves, a surf break in the sky.

We sail between the Turneffe Islands, out of sight beyond the horizon, and the mainland. If the Turneffe Islands provide protection from the swells, I neither feel nor see that protection here. The islands are too far offshore and too low. Winds blowing at twenty-five knots mock the forecast that called for fifteen knots.

At close to noon, we turn west, sailing between Water Caye and English Caye into the wide reef cut known as the Eastern Channel. Suddenly, we are inside the reef, protected, sailing toward Belize City on flat water in a well-marked channel. We share the channel with commercial traffic. A small coastal tanker passes us, headed for the city, but the channel is surrounded by islands with mangroves and palm trees and scattered wooden houses.

We have sailed into paradise.

By the middle of the nineteenth century, the Pennsylvania oil fields—the first of the world's great oil fields—were producing more crude than could be sold locally. Some of the oil, destined for use in British lamps, was loaded into 1,329 wooden barrels, and the barrels were loaded onto the sailing brig *Elizabeth Watts*. Under sail, the *Elizabeth Watts* departed from Philadelphia on November 19, 1861, taking forty-five days to deliver her cargo in London. The first petroleum to cross the ocean crossed under sail.

More than a century later, the 236-foot-long Japanese tanker *Shin Aitoku Maru*, small by modern standards, carried eleven thousand barrels of oil. Like the *Elizabeth Watts,* the tanker was powered, at least in part, by sails. When the wind blew, she flew sails stretched on metal frames, looking something like vertical wings.

And later still, in a floating design reminiscent of H. M. Fletcher's windmill wagon, the owner of a catamaran installed a wind turbine where any normal sailboat would hold a mast and sails. When the turbine spins, its power, by virtue of gears and spinning shafts, is transferred to a propeller. The propeller pushes the ship forward.

These designs, like the Flettner rotor and the giant kite of the *Beluga SkySails,* might be called innovative. They might also be called throwbacks. Either way, they have not caught on. In addition to the inertia that threatens every radical change, they face other obstacles. They add complexity. They work only when the wind blows at just the right speed or from just the right direction. They create safety risks. In the case of the catamaran, the turbine adds weight aloft and affects the vessel's stability. Stand in the wrong place, one critic said, and the spinning turbine blade could connect with an important piece of boating equipment—the skipper's skull. And another critic cut straight to the point, pulling no punches, complaining that a catamaran fitted with a wind turbine is, quite simply, ugly.

At anchor, in front of Belize City, we fly our yellow quarantine flag, waiting for the authorities. The wind blows toward the concrete city pier. We bob on the waves. If our anchor drags, we will be up against the pier in minutes, blown onto a lee shore.

Expecting the worst, I have all the cabin portholes and hatches closed. The temperature inside is well over ninety degrees. My hope is that the port authorities will not want to linger in the heat.

Four men arrive in a launch, boarding from the starboard side, storming *Rocinante*. The men are dark-skinned, two potbellied and middle-aged, one potbellied and young, and one athletic and young. They do not chat among themselves. They push past me, not waiting for an invitation before filing down into the heat belowdecks, three sitting at the galley table and one standing in the salon. Before producing their paperwork—in other words immediately—they complain about the heat.

My co-captain gives them paper towels to clear the sweat from their brows.

One of the middle-aged potbellies points a finger at me. "Didn't you see us on the pier? I was there waiting for you." This is, of course, not true. I have been watching the vacant pier ever since we dropped anchor.

Almost as a unit, clearly as a team, the men make simultaneous demands. They want passports. They want five copies of vessel documentation. They want our Mexican departure paperwork, our zarpe.

From the first middle-aged potbelly: "Don't you have a photocopier on board?" We do not.

There are forms asking about sickness on board, about fruits and vegetables, about firearms. I ask if flare guns are firearms. This triggers a passionate discussion in Belizean Creole. I understand a few words here and there—*tek* for "take," something like *gon* for "gun," *bwai* for "boy." I understand enough to know that they are making up rules as they go.

I complete a form listing the number of cans of vegetables, of fruit, of beer, and of soda that we have aboard. "Just estimate," says the second middle-aged potbelly, smiling helpfully. But by then I am already done.

They collect a fee for the forms and a fee for passport stamps and another fee for the launch. Then they demand fifty dollars each, for their services. They specify American currency. My co-captain pays them in wrinkled twenties and tens and fives. She asks for a receipt.

From the first middle-aged potbelly: "Receipt? Why you need a receipt?"

From the second middle-aged potbelly: "Your passport stamps are your receipt."

From the young potbelly: "There are no receipts."

The men collect their paperwork in neat piles and stand to leave. By my count, they have, collectively, complained about the heat no less than forty-eight times since boarding.

The man standing in the salon, the quietest of the four, waits for his comrades to file up the companionway. He tells me that he has a degree in natural resource management. "Not a bachelor's degree," he says. "There are no bachelor's degrees in natural resource management in Belize. The politicians don't want it. They only want development."

He tells me that he works only two days each week, spending the remainder of his time in a school studying maritime commerce. The school will prepare him for another school, in Sweden. "I want to climb the ladder," he says.

He asks what we do for work. "We are biologists," I tell him.

"This is the perfect boat for you, then," he says.

His comrades, now in the cockpit, growing impatient, call down to him. He starts to move toward the companionway but then turns to me again. "You know we could have charged you more," he says. "A lot more."

I nod.

"I could have charged you for the taxi fare here, thirty dollars, but I drove," he says.

I offer him thirty dollars, but he shakes his head. He climbs into the cockpit, and before I can join him, the launch thumps against the starboard side of *Rocinante*. Their ride has arrived. We are rid of them, of their complaints, of their unpredictable commands and demands, of the choreographed bedlam they created. Peace returns to *Rocinante*, bobbing on her anchor.

I lower our yellow quarantine flag and open the hatches, letting the east wind blow the heat from *Rocinante*'s cabin. We sail away from Belize City, and an hour later we anchor in the lee of an uninhabited mangrove island, comfortable for the night.

Since the birth and early life of Lewis Fry Richardson, the world has changed in many ways. Wind-powered commercial ships have all but disappeared from the world's oceans. The promise of Flettner rotors, for the most part and for now, has come and gone. Wind turbines, ashore and at sea, have proliferated. Standard radar and its specialized cousin Doppler radar have become commonplace, as have electronic computers barely imaginable in Richardson's time.

One thing that has not changed is the importance of reliable weather forecasts. The value of understanding tomorrow's weather today remains well understood. As valuable—and for some applications more valuable—is the understanding of next week's weather today, or even next month's weather. The long-range forecast, if reliable, can save lives and money. Good forecasts prevent disappointing vacations, but they also protect ships and airplanes, improve agricultural yields and reservoir management, and guide the supervision of infrastructure such as flood control gates and wind farms. A 2014 government report estimated the value of forecasts at more than thirty-one billion dollars, many times the cost of generating the forecasts.

Conclusion: Good forecasts are a good investment.

"Perhaps someday in the dim future," Richardson wrote in 1922, "it will be possible to advance the computations faster than the weather advances." Between 1950 and 1960, computers grew faster and more reliable and more available. Richardson's dim future arrived.

As forecasts improved, methods were developed to assess their accuracy. This was not simply a matter of subjective judgment. Forecasters, entrenched in mathematics, developed what became known as skill scores to quantitatively compare forecasts and actual weather. Reported as percentages, like student test scores, skill scores can indicate one hundred percent agreement for perfect forecasts to zero percent agreement for perfectly incorrect forecasts. In the 1950s, skill scores for numerical forecasts generated by early computers were stuck below the fifty percent level, marginally better than subjective forecasts. But the 1949 ENIAC forecast and its immediate successors were only the beginning. Keeping pace with computer advances, forecast models became more complex. Skill scores inched upward. By 1969, skill scores reached a passing grade, just over sixty percent.

There was reason for optimism. There was reason to believe that better computers and better input data and better models would let forecasters peer into the future infallibly. Those were heady times for people who believed in a deterministic universe, for those who thought that condition $a$ plus condition $b$ plus condition $c$ always led to condition $d$.

There was reason for optimism, but there was also Edward Lorenz.

Edward Lorenz was a mathematician trained at Dartmouth and Harvard before working as a weather forecaster for the Army Air Corps in World War II. Returning from the war, he studied meteorology at the Massachusetts Institute of Technology, where he would later become a professor.

Armed with a computer, Lorenz ran calculations in the early 1970s. One of his original goals was to show that forecasts of future weather based on the principles of physics and the application of mathematics were superior to forecasts based on the statistics of past weather. The fact that it has been cold and blustery for the past five days may be an indicator that it will be cold and blustery tomorrow, but the principles of physics and the application of mathematics would offer, he believed, a far better forecast. The work of Richardson and Charney needed cheerleaders.

Lorenz's 740-pound computer, a noisy Royal McBee LGP-30 the size of a desk, could run sixty multiplications per second. He played with a simplified weather system, with rising warm air and descending cooler air. He worked out the math to six decimal places. He entered numbers to start the model, then let the computer chug along. Based on the numbers he entered, the computer forecast the future of the simplified weather system.

Lorenz decided to try something unusual. He took output from partway through a computer run, then let the computer continue chugging along. Later, he took the output he had snatched away during the first run and reentered it, using it as input to start a second run. The computer chugged. He expected nothing interesting. He expected only that the second run, based as it was on numbers from the first run, would produce the same results. Starting with what he thought were the same numbers, he expected to end with the same numbers.

Lorenz did not get what he expected. What he thought were the same numbers going in did not give him the same numbers coming out. The second run, using numbers provided by the first run, did not do what any reasonable person would expect. It did not give the same results. The further into the future he ran his comparison of the two models, the more their output diverged. Something very strange was going on.

Lorenz's computer was sending him a message. At first, he could not believe what the computer had to say. He thought that he might have made an error, somehow entering a wrong number here and there. He considered whether something might be wrong with his computer, whether one of his machine's 113 vacuum tubes might be overheating. Then he remembered that the original calculations were being run on numbers out to six decimal places, but his new data went out to only three decimal places.

He reentered the numbers to six decimal places and ran the model again. This time it worked. Identical input gave the expected identical results.

But still, the disagreement between models run with similar but not identical numbers troubled him. It made no sense. If a banker sums a million-dollar account balance in pennies, dimes, or dollars, the banker will get a similar result each time. Not so, it seemed, with weather forecasters. Lorenz saw that one forecaster working with three decimal places and another working with six decimal places would forecast different weather.

Lorenz had discovered, or rather rediscovered, chaos theory. The present coupled with the principles of physics and the application of mathematics might forecast the future, but the approximate present coupled with the principles of physics and the application of mathematics might not.

As scientists tend to do, Lorenz wrote a paper, "Predictability: Does the Flap of a Butterfly's Wings in Brazil Set Off a Tornado in Texas?" His answer was yes, but with caveats. The first caveat: The flap of a butterfly's wings in Brazil could stop a future tornado in Texas as easily as it could set one off. The second caveat: It will never be possible to know whether the flap of a butterfly's wings in Brazil started or stopped a future tornado in Texas, or had nothing whatsoever to do with it. The third caveat: There are lots of butterflies flapping their wings out there, and lots of other actions that contribute to small changes in weather, and any one of them could, at any time, change things in any number of ways, or not.

In the fourth paragraph of the paper, he rephrased the question raised in his title: "Is the behavior of the atmosphere *unsta-*

*ble* with respect to perturbations of small amplitude?" His answer—unequivocal, emphatic, and simple—was yes.

In what came to be known as chaotic systems—and the earth's atmosphere is a chaotic system—small differences now can and do have large effects later. It is this reality that makes the butterfly's wings such an attractive analogy.

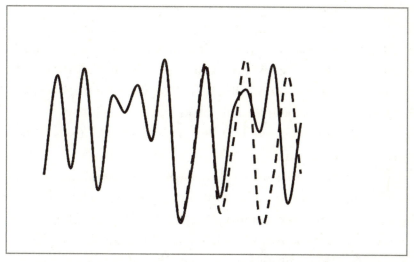

*In chaotic systems, small differences in the starting point lead to larger and unpredictable differences in output. The two lines (the dashed line is hidden by the solid line in the left half of the figure) were plotted using the same equation (the logistic equation, with* k = 3.9*), but the solid line started with the value 0.9 and the dashed line started with the value 0.90002. Time—or the number of model iterations—progresses along the horizontal axis. At first, the output appears identical, with the solid line and the dashed line overlapping, but after only thirteen iterations, the output suddenly diverges.*

The trouble with forecasting is that humans will never be able to track every flap of every butterfly's wings, or know which of those flaps might set off or stop a tornado in Texas, or which of those flaps will have nothing to do with a tornado. "If a single flap of a butterfly's wings can be instrumental in generating a tornado," Lorenz clarified, "so also can all the previous and sub-sequent flaps of its wings, as can the flaps of the wings of millions of other butterflies, not to mention the activities of innumerable more powerful creatures, including our own species."

Humans will never know every perturbation of small amplitude, or even of large amplitude, nor will they know the results of all or even most perturbations large and small, and so they will never forecast exact futures.

But Lorenz was not suggesting that forecasters give up. In keeping with the conventions of his chosen occupation, he called for more research. He also called for expanded observing systems. "It is to the ultimate purpose of making not exact forecasts," he wrote, "but the best forecasts which the atmosphere is willing to have us make."

As computers improved, cell sizes in computer models shrank, increasing resolution. The time intervals used in the models shrank from the three hours accepted by Charney and ENIAC to five minutes or less. Data from satellites and weather buoys and drifters filled in gaps.

Despite chaos theory, despite the fact that the behavior of the atmosphere is unstable with respect to small perturbations, forecasts improved. By 1992, skill scores for short-range forecasts were hitting ninety percent, up from sixty percent in less than the span of a meteorologist's career.

The problem with chaos, though, is not so much the short-range forecast. It is not what happens in an hour or two hours or six hours. Normally it is not even what happens in the next two or three days. Although a short-range wind forecast might be off by enough to inconvenience a sailor, it would not, in normal circumstances, be wildly and life-threateningly wrong. The problem with chaos is what happens next week and two weeks out. The problem with butterfly wings is that their interminable flapping may wreak havoc with long-range forecasts, the very fore-casts important to activities such as vacation planning and farming and ocean crossings on small boats.

But even there, even in the realm of the long range, con-fronted with chaos theory, forecasters have not given up.

## Chapter 9

# THE ENSEMBLE

Inside Belize's barrier reef, the wind blows but the water lies flat, the ocean waves tripped by shallow coral. We enjoy a week of easy sailing, stopping each night behind uninhabited mangrove shores. Occasionally, we talk to fishermen, some in dugout canoes and some working from launches with outboards.

We sail to the village of Placencia and drop our anchor between the shore and the little island called Placencia Caye. The island protects us from the east. Placencia itself, perched on the end of a narrow peninsula, protects us from the north. The peninsula, we have been told, was once called Punta Placencia, and whether this is true or a fabrication of the tourist industry, the name—in English, Pleasant Point—is well deserved.

*Rocinante's* neighbors include eight sailboats and a trawler. In the morning sun, we lounge in the cockpit, reading and thinking and talking, accomplishing nothing, caring not.

We watch a thirty-foot-long Belizean fishing boat approach under sail, her wooden hull dirty white, her lines sweeping from a high bow to a low stern, her decks covered with men and three brown dugout canoes. Her mast is short but her boom is long, giving her mainsail the symmetry of an equilateral triangle. She skips across the ripples with the wind directly behind her, her mainsail at a right angle off the port side, her foresail at a right angle off the starboard side, running, as some sailors say, wing and wing, cutting through clear blue water under fair skies.

With her fine lines and her oversized sails and her speed, the little Belizean fishing boat brings to mind the great clipper ships, known for their stylish shape, known for carrying an abundance of sail, and, above all else, known for their speed. Hundreds of years of commerce under sail resulted in the clippers, the ultimate sailing vessels for those in the business of moving goods from place to place as quickly as possible without the benefit of fossil fuels. But the clipper chapter was short-lived. One of the first of the clippers was the 143-foot-long *Ann McKim,* built in Baltimore in 1833. Just thirty-six years later, in 1869, the *Cutty Sark,* among the last of the clippers, was built in Scotland. The *Cutty Sark* was half again as long as the *Ann McKim,* and she was fast, reaching a top speed of just under eighteen knots, able to cover almost four hundred miles on a good day. She was built for the run between Europe and the East Indies, for moving, among other things, silver and tea and opium. But the year of her christening was also the year that the Suez Canal opened for business. The *Cutty Sark* had to compete with steam-powered vessels that could pass through the canal, a shortcut that shaved more than four thousand miles off the run between Europe and the East Indies. And those steam-powered vessels, for all their lack of grace and their noise and their billowing smoke, did not stop moving when the wind stopped.

Libraries have been filled with the volumes written about sails and sailing vessels and trade routes and the men and women propelled around the earth aided only by wind. Biographies, autobiographies, novels, and histories of life at sea abound. A cursory understanding of that literature indicates one thing above all else: on this maritime planet, humanity depended, for many years and in many places, on ships under sail. And now, but for occasional anachronisms, but for experimental wing-like sails mounted on tankers and kite-like sails cabled to freighters, but for yachts and the odd fishing boat, sails are a thing of the past.

The little Belizean fishing boat, still wing and wing, passes south of us, making for the entrance to a bay just past Punta Placencia, behind Pleasant Point. She is so fast that she leaves a wake behind. From this angle, her two sails together give her the look of a lopsided but beautiful butterfly.

Edward Lorenz did not give the name "chaos" to the strange mathematics that he pursued. The name came after his butterfly paper, in a 1975 article by Tien-Yien Li and James Yorke. The paper, called "Period Three Implies Chaos," was published in the *American Mathematical Monthly*. "In this paper," Li and Yorke wrote, "we analyze a situation in which the sequence $\{F^n (x)\}$ is non-periodic and might be called 'chaotic.'" One interpretation: The sequence $\{F^n (x)\}$ twists through a tortuous and never-repeating path. Another interpretation: The sequence $\{F^n (x)\}$ does exactly what the weather does.

Lorenz was not the first to dabble in the field that came to be known as chaos theory. Others had stumbled upon chaos. In Russia, it had been a matter of discussion for some time before Lorenz ran his simple computer models and wrote his butterfly paper. Nevertheless, Lorenz was instrumental in cornering what is sometimes called "Laplace's demon."

In 1814, Pierre-Simon Laplace—the very same Laplace whose book *Celestial Mechanics* influenced the young bumpkin William Ferrel—articulated what science wanted to believe and what, by then, most scientists already did believe. Laplace elaborated on the simple but universal principle of "if this, then that."

"We may regard the present state of the universe as the effect of its past and the cause of its future," Laplace wrote. "An intellect which at a certain moment would know all forces that set nature in motion, and all positions of all items of which nature is composed, if this intellect were also vast enough to submit these data to analysis, it would embrace in a single formula the movements of the greatest bodies of the universe and those of the tiniest atom; for such an intellect nothing would be uncertain and the future just like the past would be present before its eyes."

Laplace's demon, the demon cornered by Lorenz, was that vast and all-knowing intellect.

Laplace was right. If this, then that. Planetary motion can be predicted within reason. Rockets can reach the moon. If a planet is here now and is traveling at such and such a speed in such and such a direction, then it will reach such and such a point, at least roughly, at some specific time in the future. Its future position can be worked out accurately enough for most purposes to satisfy most human concerns. In fact, if Laplace's demon had the perfect knowledge he described, the future position of planets could be worked out precisely. Likewise, with perfect knowledge future weather could be worked out precisely. But that perfect knowledge would mean knowing where every molecule was and where every molecule was going. The number of molecules in the atmosphere, roughly speaking, could be represented by the number 1 followed by forty-four zeros.

To Laplace and his followers—that is, to most scientists and philosophers of the nineteenth and twentieth centuries—the world was deterministic, even in the approximate sense. They saw no reason that an approximate knowledge of the present would not yield an approximate but very useful knowledge of the future. Bjerknes and Richardson and Charney and Rossby and all of their friends, among others, believed that they could, in principle, forecast the weather numerically. If this, then that.

In nonchaotic systems, small differences at the beginning have no more than a small effect on outcomes. But drop two identical Popsicle sticks side by side into a stream and try to predict exactly where they will be relative to each other in two minutes, in five minutes, in ten minutes, and the only certainty is future disappointment. Try to predict the shape of a flame, or the position of a cloud, or the flow of lava, and again look forward to disappointment. Weather is not the only chaotic system in nature.

It is almost impossible to overstate the importance of Lorenz's work. He was not the namer of chaos theory, and he was not the first to see chaos in the world of mathematics, but his work brought attention to the field. He brought chaos to meteorology, and from there it spread to other fields. Its tricky machinations could be seen in electronics, in dripping faucets, in population growth. With his work, others began to see that the future can be remarkably sensitive to starting conditions, that cause may not be proportional to effect. A small cause could have a big effect.

In a world that understands chaos theory, the surprise is not that long-range weather forecasts are seldom accurate. The surprise is that they are ever right.

By the time Lorenz's ideas emerged, by the time his work began to draw attention, many meteorologists were predisposed to accept the influence of the flapping of a butterfly's wings. After a century of botched forecasts, after realizing that four-day and five-day and six-day forecasts were no more reliable than vaguely educated guesses, meteorologists were ready for an explanation, a viable excuse. Some may have accepted it with a sigh of relief. Forecasters were vindicated. What had appeared to be ineptitude was in fact inevitable.

In 2007, Lorenz was interviewed. "I do think that the meteorology community accepted it pretty well," he said of chaos theory.

But from outside the world of meteorology, and even from corners within it, Lorenz faced criticism. His mathematics were a trick, an anomaly, unrepresentative of the real world. He was a meteorologist dabbling in mathematics. He could not be right. If small changes in input lead to large changes in output, science could be thrown on its head, knocked down like a sailboat caught by a violent gust.

The wind dies by midmorning. We row ashore, pulling our dinghy onto the beach and tying her to the trunk of a palm tree. We walk to Placencia's concrete pier. There two men and a woman in sandals and shorts and broad-brimmed hats stand in a circle. They are sailors and all north of sixty years.

We introduce ourselves and then listen. They debate the forecast. "Chris calls for forty-five-knot gusts in north and central Belize by six o'clock," says one of the men, probably referring to Chris Parker, who produces forecasts for cruisers in the Caribbean.

"Chris has been wrong all week," says the woman.

"The real question," says the other man, "is whether we are in central Belize or southern Belize. Chris says it will blow in central Belize but not in southern Belize. The weather will come from the north. It could blow itself out before we see it."

"Or not," says the woman.

I look offshore, across the anchorage. The air is oppressive, thick and soggy, abandoned by the trade winds. Two new sailboats, both under power, with bare poles, make their way toward the anchorage. More, hearing the forecast, will come. They will come in from outlying islands to tuck themselves behind Punta Placencia. The anchorage, if the wind hits, could become a dangerous and expensive rat's nest of tangled anchor chains, each attached to a different boat, each under the command of an opinionated captain. The only certain qualification of the captains, the single item that certifies them as mariners, is that they all had the means to acquire a boat. The same is true for me.

One of the men on the dock asks for my opinion. Unlike them, I have not heard the forecast. I tell them, "The physicist Niels Bohr once said that prediction is difficult, especially about the future." They look at me as though I am a bit cracked, but cruisers, as a rule, are a tolerant bunch.

"I think the baseball player Yogi Berra said more or less the same thing," I add. "I'm not sure, but I think Bohr said it first."

Jule Charney, working with ENIAC on the first successful numerical forecast, relied on many others. Among the host of scientists involved, John von Neumann was critically important. He was the one who recognized that ENIAC, brought into being for military purposes, might be useful in forecasting.

Von Neumann is known for many things. First among them may be his brilliance. Second, his contributions to mathematics, including the development of game theory. Third, his role in creating the bomb that destroyed Nagasaki and his role, with Edward Teller and others, in creating the hydrogen bomb. Fourth, his witticisms, which included clever acronyms, such as MAD for mutually assured destruction and MANIAC for Mathematical Analyzer, Numerical Integrator, and Computer, an ENIAC successor. The John von Neumann important contributions list could go on and on but would certainly include his role in initiating the first successful numerical forecast on ENIAC.

Somewhere on that list, likely near the bottom, would be his interest in weaponizing weather.

"I am violently anti-Communist," he said during hearings on his nomination to a government post. "My opinions have been violently opposed to Marxism ever since I remember, and quite in particular since I had about a three-month taste of it in Hungary in 1919."

The military, he knew, could benefit from both weather forecasts and weather weapons. Military control of weather meant military control of airspace, of ground troop movements, of the high seas and the coasts.

The US military will not disclose the degree to which it has pursued weather control, but the fact that the military has been interested in this topic is indisputable. From a 1974 newspaper report on what was called Operation Popeye: "The Defense Department has acknowledged to Congress that the Air Force and Navy participated in extensive rainmaking operations in Southeast Asia from 1967 to 1972 in an attempt to slow the movement of North Vietnamese troops and supplies through the Ho Chi Minh trail network."

There were other schemes, including speculation about the use of high-energy beams supposedly capable of changing the heat balance in tornadoes, of shutting down the very engine that drives their winds. But there was also the 1978 Convention on the Prohibition of Military or Any Other Hostile Use of Environmental Modification Techniques. "Each State Party to this Convention," the treaty says, "undertakes not to engage in military or any other hostile use of environmental modification techniques having widespread, long-lasting or severe effects as the means of destruction, damage or injury to any other State Party."

The treaty, being a treaty, did not point out the reality of chaos theory. It did not point out that small causes can have big consequences. It did not say that seeding clouds in Vietnam could have unpredictable, untraceable, widespread, and long-lasting effects. Nevertheless, the treaty stopped some weather control efforts. Others likely went underground. But the militaristic desire to control the weather survived. From a US Air Force report dated August 1996: "In 2025, US aerospace forces can 'own the weather' by capitalizing on emerging technologies and focusing development of those technologies to war-fighting applications." The authors wanted rain and fog to cloak the positions of friendly forces. They wanted heavy rain to bog down enemy troops, drought to deny freshwater to enemy forces, and gray skies to impact the comfort and morale of the enemy. The authors mentioned chaos theory, but they emphasized the ability of the military. "Within the next three decades," they wrote, "the concept of weather-modification could expand to include the ability to shape weather patterns by influencing their determining factors." Their words suggest an understanding of chaos theory coupled with an underlying faith in simple determinism, in cause and effect, in "if this, then that." They were right to believe in both. Chaos theory does not dispute the predictable nature of cause and effect in the short term.

Despite common misconceptions, chaos theory is deterministic. Two identical starting points will follow exactly the same track. What chaos theory points out is that hitting two identical starting points is all but impossible. Current weather is never an exact carbon copy of previous weather, and future weather will never be an exact carbon copy of current weather. Very small differences lead to very large and unpredictable differences. It is a matter of accuracy and time horizons—the more accurate the input, and the shorter the time horizon, the more predicable the outcomes. Provided that the longer-term unintended consequences are of no consequence, provided that one is worried only about victory today and victory at all costs, provided that one is willing to ignore both treaties and common sense, weather weaponization makes perfect sense.

"It is just as foolish to complain that people are selfish and treacherous," von Neumann once said, "as it is to complain that the magnetic field does not increase unless the electric field has a curl. Both are laws of nature."

And this: "If people do not believe that mathematics is simple, it is only because they do not realize how complicated life is."

My co-captain and I wander somewhat aimlessly down a sandy lane in Placencia. We follow signs to the Hokey Pokey water taxi dock, where passengers board to cross from the peninsula of Placencia to the larger and less touristic combined towns of Independence and Mango Creek, where bananas are loaded onto ships and sent north, to North America. The water taxi crosses a protected bay.

I eye the bay as a potential storm anchorage. A man tells me that manatees and crocodiles live in the bay. The depths are shallow. We walk on, past bars and real estate offices and small grocery stores with dust-covered canned goods visible through open doors. The air remains oppressively still.

We stop at an outdoor restaurant with a single employee, who also happens to be the owner. She is dark-skinned, smiling, heavy, with an ample and ill-concealed bosom. She is, without doubt, Garifuna, her blood a delightful mix of African and Carib and Taino. It was her Taino ancestors who passed the word *hurakán* to Columbus.

She sets our table with cheap steel spoons and forks, neatly placed on top of clean cloth napkins. She suggests chicken with beans and rice and beer, but she cannot sell beer. We have to get the beer across the street, from a tiny grocery store.

I ask for the name of her restaurant. She says it has no name. Her own name is Mary. I suggest calling the place Mary's No-Name Restaurant. She sees this as a rich joke. "Mary's No-Name Restaurant," she says, smiling, then repeating it twice and laughing each time.

When we finish our meal, she picks up our dishes. "Going to be cold tonight," she says, "and wind is coming. Cold and rain and wind."

She does not say how she knows this, and I do not ask, but I note the certainty of her words, as though the future were as factual as the past, as though the possibility of anything but cold and rain and wind was not worthy of discussion at all.

Down the sandy road from Mary's No-Name Restaurant sits a shack, twelve feet long and twelve feet wide, advertising haircuts. In its shadowy innards sits a single chair. The barber pauses his clipping to say that I should come back later.

Across the street I surf the Internet in a coffee shop for tourists, comparing forecasts. One, with a rainbow-colored header and advertising banners for hotels and restaurants and tours, predicts nothing but sunny days and star-filled nights in perpetuity. Four others, more helpful, are less optimistic. They show wind coming our way, a vicious blue norther that swept through Texas and across the Gulf, still full of energy and still moving, pushing deep into the trade winds belt.

No two of the forecasts are identical. They call for wind speeds from twenty to forty knots. They do not mention what I already know — that the pressure pushing against a boat increases with the square of the wind's speed. That is, a steady forty-knot wind exerts four times the pressure, not twice the pressure, of a steady twenty-knot wind. Put another way, a boat in a forty-knot wind needs an anchor four times more capable than one that would suffice in a twenty-knot wind. Put yet another way, loosely speaking, a forty-knot wind is four times more likely than a twenty-knot wind to turn the anchorage into a tangled mass of fouled tackle with damaged boats commanded by swearing captains.

There is another difference, too, between the forecasts. One shows the norther blowing right through Placencia and continuing south to Guatemala. Another shows it petering out just south of Placencia, near Monkey River. A third shows it dying in Dangriga, to the north. PassageWeather, one of the few websites to explain where its forecasts come from and to name the numerical models on which it relies, shows it blowing hard off-shore but easing to below twenty knots onshore and maybe even in our anchorage.

It seems reasonable to think that the truth will lie somewhere in the middle of the combined forecasts, or at least in the range of what they predict.

I recross the sandy street and sit in the single chair while the barber clips away. And while he clips, he chatters. I listen. He learned to cut hair from his cousin. He came here to start a business but will not live in Placencia for long. "Too expensive, man," he confides. And people make fun of his shack of a shop. But he stays busy. A salon would do well, he thinks. "The tourist women love to get their hair done," he says. "But I only do men's hair. And women want women to cut their hair."

He says nothing about the cold or rain or wind, and if he has ever heard of chaos theory, he makes no mention of it.

Much has been made of chaos theory, and rightfully so, but chaos theory does not entirely ban forecasting. Chaos theory says, in one sense, that detailed forecasts will be more prone to errors than generalized forecasts, and that long-range forecasts will be more prone to errors than short-range forecasts. It is possible within the realm of chaos theory to predict tomorrow's weather with enough detail and enough certainty for the forecast to be of use. It is well within the grasp of forecasters to see tomorrow's rain and wind for a region or a city, to see how it will change from one hour to the next with reasonable accuracy. It is also possible to make sweeping generalizations far into the future. No one needs a mathematical model to predict warm summers and cold winters. But the best-trained forecaster cannot say whether or not it will rain and how hard the wind will blow for a region or a city on a particular day next summer. And the best-trained forecaster cannot say how the seeding of clouds on a battlefield might affect next year's hurricane season half a world away. This limitation is not something that can be tweaked out of the system. It is part of the system. Small differences in initial conditions lead to large and unforeseeable differences in future conditions.

But Lorenz saw more than that.

In the popular imagination, chaos theory focuses on disorder. But the popular imagination fails to see the patterns that Lorenz saw. Chaos theory, it turns out, does not allow unfettered chaos.

In his early numerical experiments, Lorenz did not look at anything resembling actual weather. Limited by the abilities of his clunky Royal McBee LGP-30, he played with a model of simple convection in a closed system. He modeled something resembling water slowly warming in a pan. He had his computer jump ahead in six-hour increments, or time steps, computing the convective flows over time. What he found at first is what is most often related in discussions of chaos theory. If he ran his model twice, slightly changing the input data between runs, the output for two runs would agree well enough in the fifth and tenth and twentieth six-hour increments, but then, somewhere out there in the future, their output would drift apart. The solutions, which appeared identical at first, would diverge over time.

If he had stopped there, he could have explained chaos theory in a single sentence: Systems are chaotic when small differences in starting conditions result in large differences over time. But Lorenz kept going. He ran models again and again.

For simplicity's sake, think of Lorenz running a model five times, each time with slightly different input and each time looking at a single point of output in the distant future—say, four months ahead. The five points were all over the map. His results appeared random.

But instead of looking at a single point in the future, Lorenz looked at multiple points in the future. He plotted the points.

What he found was nothing short of remarkable, times two. It was remarkable for single-model runs, and it was remarkable for the five-model runs. The solutions for a single-model run, when plotted in three dimensions, with each plotted point representing a step forward in time, formed a strange shape. The lines making up that shape—the lines representing the data—spiraled. They formed two intricate lobes, strangely beautiful, strangely resembling, when viewed from the right angle, the wings of a butterfly. But remarkably, the lines never crossed. The solution for a single input never settled into a cycle, never repeated itself. Even in this simple mathematical system, the solution, like the weather, remained infinitely variable.

Even more remarkable was the comparison of the runs. Each run, with slightly different starting points, resulted in a spiraling, twisted figure eight. The spiraling figure eights, laid one on top of the other, appeared almost identical to one another for a time, each new set of lines lying on top of one beneath it. But as time wore on, the overlap broke down. The spiraling figure eights continued to appear similar, each a slightly twisted pair of adjacent spirals, but each unique.

To a scientist, to a lover of numbers, what was more beautiful than the appearance of the plotted data was the meaning of the plotted data. The plotted data showed a pattern of possible solutions. When the input changed slightly, a researcher looking at only a handful of points in the output would see randomness, or what appeared to be randomness, but a researcher looking at thousands of points would see a pattern. Small changes in initial conditions led to large changes in the future, but not random changes. Not wild changes. Not changes ungoverned and unbounded. There was a pattern to the possible solutions, just as there are patterns to weather. The changes might be all over the map, but they were always on the map.

Lorenz's patterns converged around what mathematicians call a strange attractor. An iron pendulum, modeled mathematically or swinging in reality, influenced by friction, moves back and forth. An attractor—a normal attractor, not a strange attractor— draws the pendulum to zero motion—that is, hanging more or less straight down, regardless of its starting point. The attractor that Lorenz found, the attractor that forecasters have become familiar with, has strange characteristics. Take that same iron pendulum but magnetize it and swing it over a bed of magnets, each capable of influencing its path. Set in motion once, it will follow a seemingly erratic path. Set in motion a second time, with the second starting position just slightly different from the first, and it will again swing in a seemingly erratic path, but not the same erratic path it followed the first time. Set in motion a third time, again with the starting position just slightly different from its predecessors, and the pendulum will find yet another path. Set this strange pendulum in motion a thousand times, with each start ever so slightly different from all the others, and each time the path will vary. But over time a pattern will emerge. The paths are not the same, but they are not entirely different. Paths outside what might be called the strange attractor's space never occur. It is not just that the pendulum swings stay within a certain circumference, but they stay within certain lobes. There are zones where the pendulum swings and zones where it does not.

When Lorenz ran his Royal McBee LGP-30, he plotted motion around a strange attractor and saw a pattern that never repeated itself. Bear in mind that his pattern had two lobes and that, when viewed from the right angle, the lobes looked, coincidentally, like butterfly wings. Imagine starting with two adjacent points on one of the lobes, then putting the model in motion. The two points would travel together for a while but then part ways, ultimately winding up unpredictably far apart, but always somewhere on a plot resembling the two-lobed spiral, the butterfly wings. In the near term, while the points traveled side by side, their future could be seen. In the long term, vague generalizations could be made with confidence.

Forget about Lorenz, with his simple models and his Royal McBee LGP-30. Imagine instead three forecasters, each using the same mathematical model but with slightly different initial conditions, each armed with a supercomputer. Say that all three start with the same set of barometric pressures, but one of the pressures comes from a Voluntary Observing Ship that is known for inaccuracy. One forecaster ignores that barometric pressure altogether, relying entirely on the other pressures in the set. One forecaster makes what seems to be a reasonable correction to the reported pressure and inserts the value into the model along with the other pressures. The third forecaster keeps the reported pressure. The three forecasters expect to generate similar forecasts, and for the first day they do. But by the third day their forecasts begin to drift apart. On the tenth day, the first predicts a gale, the second predicts a thirty-knot wind, and the third predicts a light breeze.

Imagine the three forecasters looking for something regular, for signs of the clockwork universe that they know, as scientists, must exist. Imagine each feeling like the other had somehow botched the calculations. Imagine them, collectively, feeling rather clownish. Imagine them, in their embarrassment, not submitting their results to erudite journals. Imagine them moving on to other fields, to endeavors that made sense, to something more cooperative than weather.

But now imagine five hundred forecasters, again each with slightly different initial conditions. Imagine them—a stretch, for anyone who knows scientists—working together. Imagine them seeing a pattern emerge. All of their solutions fall somewhere within that pattern. None of them can guess where their solutions will fall within the pattern, but none of them generate solutions outside the pattern.

Wild weather outside that allowed by the rules that govern nature never occurs, but the exact same weather never occurs twice. The spiraling butterfly wings converging toward the attractor—representing certain physical realities of the atmosphere—would not allow for winds of five hundred miles per hour or for completely still air any more than they would allow two sets of starting conditions, no matter how similar, to follow identical paths into the future.

Armed with chaos theory, with an understanding of deterministic chaos, forecasters understood why their long-range predictions seemed no better than wild guesses. They understood that Laplace's demon, if it knew everything perfectly well with infinite precision, would know the future. But they also knew that they were not Laplace's demon. They knew that they could never know everything perfectly well with infinite precision. That would require knowing where every molecule of air sat at any given moment and where it was going.

Deterministic chaos—that is, chaos theory—was, at best, difficult to explain to a public that simply wanted to hear about next week's weather. Forecasters understood why they could never get the long-range forecast just right, but still, they could never get it right. They might be able to invoke chaos theory as an explanation for their failed forecasts, but they longed for something more useful. They were sick of being the brunt of jokes.

And so they turned to something that became known as ensemble forecasting.

We untie our dinghy from the palm tree, carry it across the beach, and row out to *Rocinante*. Back aboard, we discuss and then dismiss the possibility of breaking out our heavy storm anchor, which is stowed below in three pieces that have to be individually hauled on deck and bolted together. Instead, we break out our book on anchoring. We learn nothing new.

The wind, if it comes, will come from the northwest. Where we sit now, a northwest wind will carry us onto Placencia Caye, the little island that protects this anchorage from the east. We decide to move.

I think of the ships torn from their anchorage and cast ashore in the Crimea in 1854 — the *Rip Van Winkle,* the *Progress,* the *Wild Wave,* the *Kenilworth,* the *Wanderer,* and the *Prince* — all pounded and splintered on rocks and sand, all lost. I think of the yacht anchored off the Dry Tortugas, thrown ashore. I think of the boats stranded beneath the mangroves at Isla Mujeres.

We start *Rocinante's* diesel, and with my co-captain at the helm, I haul the anchor aboard. My co-captain finds a new spot, well clear of the other boats. I lower the anchor. Our storm anchor remains below, in three pieces, but we have let out more anchor line, increasing our scope and with it our anchor's holding power. My co-captain puts *Rocinante* in reverse, backing away until the anchor line pulls tight, and she holds it there for a moment, burying the anchor's flukes into the bottom.

In our new position, we are well clear of the crowded anchorage. We will avoid the rat's nest, if it happens. And if our anchor drags, we will be pushed past Placencia Caye, swept out toward open water and not onto the little island's beach. The only obstacle in sight is a single anchored sailboat, a tiny and unlikely target a half mile off.

Relaxing in the cockpit, I pick up John Caldwell's *Desperate Voyage.* Sailing in a boat named *Pagan* just after World War II, alone in the Pacific on a voyage to rejoin his young wife in Australia, Caldwell wrote of being demasted in a storm. "What I saw," he wrote, "was a monument to chaos." He was not writing of mathematical models.

"A dismasted boat," Caldwell wrote, "is a naked sight; but one in the condition that *Pagan* was in—battered and bruised, limping as if wounded—is a heartbreaking sight for a seaman."

Later, crawling across the ocean in his wrecked boat, propelled by a jury-rigged sail, he gradually starved. Before his eventual rescue, he ate every delicacy aboard, including a leather shoe, growth scraped from the bottom of the boat, and oil drained from the nonfunctioning engine.

I look up from my book, watching for the wind. It is five o'clock in the afternoon. The air, unconcerned about my concerns, hangs silent and inert. If the wind will come at all, it should arrive soon.

Forecasters, faced with the impossible challenge of chaos, came up with a practical solution. Instead of running a model just once, they ran it again and again, changing the input here and there with each run and plotting the results. They called the approach ensemble forecasting.

Ensemble forecasting offered a glimpse into the future, despite the chaos. If the models gave two or three or four days of similar results, the forecasters knew they had a reasonable forecast for two or three or four days. If they ran the model a hundred different ways and got back a hundred wildly different forecasts for day three, they knew they were in trouble.

Forecasters played with this approach as early as 1969. It is a practical solution, simple enough for anyone to understand in principle. But it required tremendous computer speed. And it required reasonable changes between model runs—not just random changes, not just made-up numbers, but changes that could represent the real world.

The forecasters tinkered, making incremental improvements in every way they could. Some of the improvements came in the form of more real-world data from weather buoys and drifters. Some came from more and better satellites. Forecasters would never have the data available to Laplace's demon, but every new station reporting in gave them something, especially when it came from those parts of the world that were, for one reason or another, empty—places like the far north and high mountains and the distant Pacific outside the shipping routes. All of the data went into the models.

Some of the improvements came in the form of better models. In the early days, the models were geographically limited. A forecaster, burdened by limited computer resources, looked at a region. The forecaster's models had borders. At borders, the forecaster had to make assumptions.

As computer speed doubled and doubled again and doubled again, certain simplifying assumptions needed for computers like ENIAC and the Royal McBee LGP-30 were thrown aside. Models encompassing the entire earth were needed, and they were built, solving problems associated with artificial borders that bounded regional models. Methods of combining regional models with global models—nesting, as the technique is called—followed.

Along with improved models and faster computers, ensemble forecasting advanced. At the simplest stage, an ensemble forecaster worked with a single model, changing only the input values, the initial data. But before long, the forecaster could run ensemble forecasts on more than one model, producing what became known as a multimodel ensemble forecast. If the forecaster adjusts those models to account for their shortcomings, the forecaster produces what became known as a superensemble forecast. If the forecaster's ambitions remain unsatisfied, models of ocean currents and ocean temperatures can be combined with atmospheric models. Now the forecaster works in the realm of the hyper-ensemble, an ensemble of assembled ensembles.

Edward Lorenz himself, in the early days of chaos theory, played with different weather models. He looked at models developed by others. He searched for doubling rates of errors. He did not completely explain his thinking. He did not, for example, say what sorts of errors concerned him. Were they errors analogous to rain versus shine, or still air versus hurricane-strength winds, or snow versus sleet? The details were not relevant. His conclusions were optimistic. In looking at a colleague's complex model, a model with many factors at play, he noted with some skepticism that the number of typical errors doubled every five days. "If it really requires as long as five days for typical errors to double," he wrote in 1964, "moderately good forecasts as much as two weeks in advance may some day become a reality."

By 2008, the year Edward Lorenz passed away, forecasters had stepped closer to that reality. Forecasters armed with faster computers, better real-world data, and ensembles could predict weather up to two weeks into the future. The forecasts might not be certain, but the forecaster would have some idea of the odds, or the probability of the forecast being close to something that would actually occur. The forecaster might know that the ensemble shows almost no differences between models, or that it shows wild differences, and now it becomes a matter of somehow conveying that uncertainty to a public that, despite the challenges, expects nothing less than perfect skill scores and wants next week's weather today.

At five thirty in the evening, the sun hangs low over the village of Placencia, accentuating the heat haze that has settled over Belize. There are no clouds in sight. There is no thunder. The storm has not reached our position. It is a fish storm, someone else's problem, blowing out at sea but not reaching our anchorage.

It is too hot to cook below. I grill vegetables and fish in the cockpit.

At six, now on the darkening side of twilight, we sit with our dinner and a glass of rum in the cockpit.

I am two bites into my fish when the wind hits. In seconds, the wind speed goes from nothing to forty knots.

I throw the rums overboard and grab the full dinner plates, tossing them forward and away from the helm. *Rocinante* swings hard on her anchor. Right away I see our mistake. Despite all of our reading, or maybe because of it, we set the anchor without thinking about the wind's direction. In this wind, *Rocinante* will pull her anchor backward, tearing its flukes from the mud.

I start the engine. The anchor line pulls tight.

Rain falls, scattered heavy drops for a brief moment, and then a downpour. Raindrops propelled by a forty-knot wind feel something like pellets fired from an air rifle.

It is not one or two or three drops, but a stream of drops, no less than would be expected from a showerhead in an expensive hotel, except that the drops are cold and forceful and they hit sideways.

The anchor line is tight, but through the rain I make out the vague shapes of other boats in the anchorage north of us. We are leaving them behind. We are moving. We cut a troubling wake, backward, through the water.

I put *Rocinante* in gear, trying to ease pressure off the anchor. The *Prince*, the pride of the British transport service in 1854, ran her steam engine to ease the tension on her anchor chains, but she eventually succumbed to the wind, snapping her chains, and ended her days on the rocks of the Crimea. The *Royal Charter* followed a similar path to her fate on the rocks of the Welsh coast. I do not intend to imitate the *Prince* or follow the *Royal Charter*.

My co-captain, on the bow, watches our anchor line. Steaming too far forward, entangling the anchor line in the propeller, would be less than ideal. Even slacking the anchor line would be less than ideal. A tight line might reset the anchor, allowing it to dig in.

My co-captain yells from the bow, forty feet away. I hear her yelling but cannot make out her words. Her words disappear against the backdrop of wind and rain.

I have our electronics on, tracking our location. We move at almost a knot, backward, with our anchor following along, apparently skipping over the bottom, useless. I have to wipe the rain away from the electronic screen. On our current course, we will clear Placencia Caye.

On the anemometer, *Rocinante* registers gusts of forty-seven knots, to Beaufort in 1831 a strong gale gusting to the edge of a whole gale. The gusts come one after another, each at first a surprise, but soon enough just an expected undercurrent of the steady wind.

In the thickening darkness behind *Rocinante*, a shape looms in the rain. It is the lone sailboat that had, when we anchored, seemed so far off. Now it approaches us, or we approach it, with alarming speed. I increase the throttle, hoping to slow down.

We are less than two boat lengths from her. She is hard to see through the rain, but it is clear that no one is on deck. She is unlit. She appears to be unmanned.

I am not, at this particular moment, in complete control of *Rocinante*.

The pelting rain, stinging, becomes painful.

I turn the helm to port. *Rocinante* moves away from the other boat. I run forward for a moment, trying to see our anchor line. It runs parallel to the anchor line of the other boat. It is possible that the lines are crossed, that one anchor is already tangled in the other.

All we can do is maintain our position, working the engine against the wind. Full night is upon us now.

Another fifteen minutes pass, and the wind seems to waver, slowing for a moment or two, teasing us. During a lull—and by a lull I mean a period when the wind drops a knot or two, when the gusts ease up—I break out my co-captain's foul-weather coat. In another lull, I run the foul-weather coat forward to my co-captain. I am cold and wet and pumped full of adrenaline. My co-captain is not even fazed. If anything, she enjoys the sport of it, watching the anchor line in a flashlight beam through the sideways rain, one hand on the rail, her wet hair plastered to her head.

Even standing side by side, even in the lull, we have to yell to be heard over the wind and rain. On my insistence, she takes her foul-weather coat.

In another lull, I break out my own foul-weather gear.

The lulls grow more frequent. I shiver.

An hour later I remain at the helm, but I no longer have to throttle up against the wind. The blow has dropped to twenty knots. Our anchor holds. We swing between one and two boat lengths from our dark neighbor.

In the rain, we debate. If we stay where we are, we will eventually swing into the other boat. We have to move. We consider cutting our anchor loose and re-anchoring elsewhere with the storm anchor. We consider trying to raise our anchor, but worry about pulling it halfway up only to find it entangled and unrecoverable. We consider radioing for assistance.

In the end, we decide to pick up our anchor. If we are entangled with our neighbor, we can cut it loose.

I throttle forward while my co-captain brings in the line. As the line comes in, we hope to see if we are entangled. It looks like the two lines cross, but we see that steaming to port might solve the problem. This seems to make matters worse, so we steam to starboard. We go back and forth. With all but seventy-five feet of anchor line retrieved, it remains impossible to know if we are entangled.

We steam farther forward. With fifty feet of line in the water, we see that we are not entangled at all. Somehow, we have moved just south of our dark neighbor. Our anchor could not have missed hers by more than thirty feet.

We recover our anchor. As it comes out of the water, I see a short length of scrap cable wedged in its flukes, trash from the bottom rendering the anchor all but inoperative. On deck, I pull the cable clear.

We steam toward shore into the twenty-knot breeze. We drop the anchor anew. We back down on it, digging its flukes into soft mud. The anchor line pulls taut. *Rocinante* swings and then halts, her backward movement arrested.

We take turns on anchor watch, one pretending to sleep while the other eyes our location.

By three in the morning, the blow dies to fifteen knots. By four in the morning, it seldom gusts past ten knots. Our first gale is behind us. To experienced cruising sailors, it would have been nothing, or at most an inconvenience. To us, it was humbling and exhilarating and educational, a fine lesson in sea room and anchoring, but also in the proper way to interpret forecasts.

In our stupidity—in my stupidity—we could have lost *Rocinante* altogether. But for luck, we could have lost our beautiful boat.

## Chapter 10

# AFLOAT IN THE CANDLE'S LIGHT

*Rocinante* lies secure, tethered to a dock twenty miles inland on the Río Dulce, Guatemala's Sweet River. Placencia lies far to the north. Here, we share this tiny marina with seasoned sailors, with boats flying Danish and Dutch and French and Belgian flags, with vessels that have crossed the Atlantic. We may feel like sailors now, having crossed the Gulf and coasted along Mexico and Belize, but we remain novices.

I sit on the dock's edge under a bright sun with my feet dangling over the river's brown water. The morning loiters without a hint of wind, as if the air died during the night, but I know it is not dead. Air is never dead.

The Río Dulce is wide here, lake-like, with no visible current. Howler monkeys, the loudest of terrestrial mammals, call from the forest behind the dock. Somewhere in the shadows, a male perched high on a branch screams for a female. His demonic throaty bellow carries for miles through the still air.

Although it happened only a few days ago, the fearful night of the storm is already but one of many memories. It is also an embarrassment. In the daylight following the blow, we were the talk of the harbor, the only boat to have moved in the anchorage. We resolved to add more chain to our anchor and to switch to a bigger anchor at the first opportunity, and we sailed on.

During the short passage from Placencia to Guatemala, the

ocean winds were as fickle as the winds of the northern Gulf of Mexico. The reliable trade winds I expected, the trades plotted on Halley's maps, let us down. Weather models and weather buoys showed them blowing out in the Caribbean, well clear of land, but near the coast they were overwhelmed by the shore effect, by localized big whirls and little whirls. We encountered light breezes that changed direction every few minutes. *Rocinante* wallowed in the waves. Her booms banged. Her sails slapped. We ran her motor.

As is common where rivers meet the sea, a sandbar stands guard at the mouth of the Río Dulce. The trick is to cross the one hundred feet of shallow water at high tide, plowing ahead with confidence, knowing but not caring that the bottom of the keel will flirt with the top of the mud, that the two might occasionally kiss, but that their relationship, if all goes well, will progress no further. The shallow crossing, so close to the end of our first real voyage, reminded us of our departure, a departure hampered by shallow water, our keel plowing through the muddy bottom near our slip on Galveston Bay, the water of the bay temporarily lower than normal, stolen by a north wind.

Despite the sandbar's reputation, we crossed without mishap to anchor in front of the tiny port town of Livingston. We rowed our dinghy in to meet an agent who took care of the paperwork allowing entry into Guatemala, a seamless process having nothing in common with the entry process in Belize. We strolled up and down the busy narrow streets, admiring the razor wire and bars that protect residences and stores, seeing all of Livingston within thirty minutes. We stopped for chicken and tortillas. We talked to a Mayan schoolteacher who walks one hour each way to the jungle village schoolhouse where he works.

We left Livingston soon after noon. Prudent sailors do not stay in Livingston after dark.

Leaving, heading upriver, the wind came from behind. We

set our foresail and let the air propel us against a gentle current. As the river twisted and turned in broad meanders, the wind turned, too. Magically, unbelievably, as we changed course the wind followed, not perfectly, but well enough to allow progress under sail. It was as if the moving air rode along a track, confined by the same channel that held the river's water.

We sailed under towering limestone cliffs with scattered vertical patches of pale gray and white stone peeking through the dark green of jungle growth and shadows. We sailed past hot springs. We sailed past Mayan fishermen throwing nets from dugout canoes. We sailed past launches with waving tourists, headed, like us, from Livingston to Fronteras, upriver. We sailed where the river widens into a lake of sorts, and where houses dot the shoreline and the noise of trucks crossing a bridge spills out across the water's surface.

We—*Rocinante* and her crew of two—sailed, mostly silent, speechless, simply enjoying life by virtue of a favorable wind.

I could not understand why, on this short inland leg of our voyage, the wind was so cooperative. And yet happily, gloriously, it was.

There are, in the annals of human thought about moving air, two repeated themes. There is the theme of measurement, of thoroughly documenting the here and now, and there is the theme of theory, of making sense of nature. FitzRoy recognized these themes, as did Bjerknes and Richardson and Lorenz. For those who think about wind and weather, these themes cannot be missed.

Before all else, someone has to look out the window. Someone has to measure and record what has come to be known as initial conditions. FitzRoy worked with fifteen stations, all on land, linked by telegraph. Today, ships and moored buoys and drifters transmit data. Passenger planes transmit data. Satellites transmit data.

Interested in current conditions at Vostok Station, Antarctica, usually described as the coldest place on earth? Right now, as I write, the barometric pressure is 1,028 millibars, and today's temperature touched a low of negative eighty-three degrees Fahrenheit. Want the water temperature in the open ocean south of Puerto Rico and north of Venezuela? According to an unmanned ocean glider, a kind of a robot submarine, it is eighty degrees Fahrenheit down to a depth of about three hundred feet, then colder down to three thousand feet. For a broader view of, say, most of the planet's atmosphere, turn to Geostationary Operational Environmental Satellite 13, or GOES-13, a seven-thousand-pound orbiter locked in at a point somewhere above Colombia, and its sister, GOES-15, locked in above a lonely piece of the Pacific Ocean. Each sees, at any one moment, roughly a third of the earth. For a view of the poles, turn to any one of the three-thousand-pound polar-orbiting satellites.

Still, the data are inadequate. Better initial conditions will yield better forecasts. Researchers, as tireless as the wind itself, come up with new ways to collect data.

An organization called PressureNet harvests barometric pressures from smartphones—not just the phones of ten or fifteen software developers and friends, but the phones of a multitude of scattered strangers. Anyone with a smartphone equipped with a pressure sensor—a surprisingly common feature—can sign up. From PressureNet's website, in bold banners: "Current weather infrastructure is too expensive and produces inaccurate forecasts," and "Low-resolution surface observations seed models with incomplete data," and, alluding to chaos theory, "Forecast errors cascade through models as forecast lengths increase." In a scheme that would have delighted the likes of FitzRoy, Le Verrier, Bjerknes, Richardson, and Lorenz, PressureNet crowdsources weather data from the masses.

Working as a team, my co-captain and I adjust the lines on *Rocinante*, tightening a bow line, loosening a spring line, recentering our beautiful boat between the pilings that will hold her in place while we, her crew, are gone. We have only one more day before we fly away, leaving her here at the dock in this tiny marina to fend for herself while we return, for a time, to what some consider to be a more normal life.

The sun rides higher in the sky now, and beneath it the wide river stands greasy calm, almost steaming under the oppressive heat. I sweat.

A German woman, short and stout and tan, waddles over from a neighboring boat, a catamaran. "We saw you in Placencia," she says, struggling with English but smiling broadly. "Your anchor—it slipped, no?"

To avoid the heat and the German, I move to the shade under the thatched roof of a *palapa,* an open-sided shelter. What this day needs is a breeze, moving air.

It is before noon, too early for a cold beer, even here in Guatemala. So instead I sit in the shade, looking over a collection of archived forecasts and weather reports from sea buoys. The mismatch between the archived forecasts and the actual conditions for some regions is striking. The seven-day forecast and the three-day update and the one-day update often disagree with one another, each adjusted as time passes to account for new initial conditions, each updated as the forecasting window grows shorter. As expected, the short-range forecasts are more likely than the long-range forecasts to agree with the actual sea buoy weather reports. But the sea buoy weather reports, at times, appear entirely ignorant of the forecasts.

Forty-three days ago, we set out to find the reliable trades, only to find that they are not so reliable after all, at least not at this time of year, not in this latitude, not close to shore. And while I learned about the winds themselves, I learned as much or more about the history of the science behind them. Knowledge of the two things—the winds and the science behind them—lies hopelessly intertwined, like tangled anchor lines. I also learned to limit my faith in forecasts and to get updates as often as possible. I learned to look at more than one forecast and to combine them, creating, in my mind, my own ensemble. In Placencia, surrounded by darkness and pelleted by sideways rain, I learned to interpret carefully, to assume the worst.

In the end, those foolish enough to sail on the sea, under way well out of sight of land or anchored near a hazard-ridden shore, must go prepared not only with the best forecasts but also with the equipment and skills that might be needed if the forecast is wrong. They must go, too, with the right mind-set. Joshua Slocum, nearing the end of his single-handed voyage around the world in 1898, wrote, "A strong hope mastered my fears."

If the hunger for improved initial data is insatiable, so, too, is the hunger for improved theory. The push to improve mathematical models, the kinds launched by Richardson, continues. Richardson's basic approach to modeling, using the finite difference method across a three-dimensional grid, remains in play, but it is supplemented by newer approaches, such as the finite element method and the spectral method, similar in principle to Richardson's approach but with different mathematics.

There is, too, a push to improve what atmospheric scientists call "parameterization." Certain aspects of weather occur at scales too small to be captured by the grid at the heart of mathematical models. Even commonly occurring clouds—the puffs of cottony cumulus clouds seen in summer skies, for example— are far smaller than the cells used in mathematical forecasting models, so modelers have to be creative. The same is true of hills and valleys and buildings. Scientists have to represent the clouds and the hills and the valleys and the buildings by averaging, in a sense, what might happen throughout the cell. They have to apply what begins as subjective judgment and alter it with experience from the real world.

Aside from aspects of weather that occur at scales smaller than grid cells, there are others that are too complex to model, too poorly understood to be described numerically. And so again, atmospheric scientists turn to parameterization. They do what they can, and models improve with their efforts. As models evolve, the demand for computer power increases. ENIAC's mathematics can and have been programmed into a cell phone, where they were renamed PHONIAC, for Portable Hand-Operated Numerical Integrator and Computer. What could be done by ENIAC in twenty-four hours can be done by PHONIAC in less than a second. But the modern mathematical models, the great-grandchildren of Richardson's efforts, the grandchildren of ENIAC, require ever more computer power. Government computers in Virginia and Florida imbibe more than two hundred million weather observations each day and subject those observations to between one thousand trillion and five thousand trillion operations in a single second. And yet the computers are still too slow.

Forecasters—the ones working the computers, the ones tweaking the mathematics, the ones taking cutting-edge science and expressing it mathematically—worry about inaccuracies in their forecasts. They worry just as FitzRoy once worried. And, like FitzRoy, they face criticism from uninformed critics and from those whose criticism might be shaded by conflicting interests. But today's forecasters talk in terms that would have left Fitz-Roy hard aground. They complain about software bugs and model biases. They worry about computer power. The capacity to do what they want to do, what they hope to do, even with computers capable of running between one thousand trillion and five thousand trillion operations in a single second, is, quite simply, inadequate. Despite all the satellites and drifters and airplanes and buoys, they talk of sparse initial data, especially upper-atmosphere measurements, especially the sort of data that come from high-altitude balloon releases. In the United States, they worry about being outdone by Europeans, whose models, at times, offer advantages.

In the book describing his botched mathematical forecast, Lewis Fry Richardson offered words that could be the mantra of every forecaster, that could be printed on T-shirts for the next World Meteorological Congress. "The scheme is complicated," Richardson wrote, "because the atmosphere is complicated."

Around two in the afternoon, a breeze appears. I wander around the dock with a handheld anemometer, measuring wind speeds. At the water's edge, I read six knots. In the shadow of a pretty red boat, I read three knots. Next to the German's catamaran, I read two knots. In the palapa, I read two knots again. Standing on top of *Rocinante*'s boom, reaching as high as I can, I read six knots. Even in a light breeze, there are whirls, big and little, wind curving around the edges of structures, sped by the heat of a hot dock above cooler water, slowed by friction, confused. And the speed of the moving air—the speed that I measure—changes from one minute to the next, or, more accurately, from one instant to the next. My little anemometer, like all anemometers, gives me a short-term average of sorts for one location while suggesting an accuracy that does not reflect reality. Rear Admiral Beaufort, with his broad categories, lived in a more certain world. He would not have walked around with an anemometer. He would have simply noted "light air," or maybe "light air tending toward a light breeze," and been done with it.

Soon our lives will change. We will return to the routine of normal jobs in normal suburbia, of e-mails and telephones, of commuting by car and bicycle instead of by dinghy. We will no longer live so close to the elements. We will no longer stir from slumber already wondering what the wind is doing, and we will no longer have to sleep in shifts. I will no longer be justified—if I ever was—in wandering around in public with a handheld anemometer. For a time, I will have to walk away from my obsession with the wind. But I will walk away a changed person, unable to ignore the slightest of breezes, or to view clouds without looking for evidence of atmospheric waves, or to hear a forecast without thinking of Lewis Fry Richardson penciling calculations to the sound of distant artillery. I will no longer be able to think of the atmosphere as anything less than a pulsing fluid skin, bulging here and contracting there, full of vortexes small and large, scarred by jet streams, alive. After forty-three days, I will leave *Rocinante* entirely unsuited for the routines of what passes in this world for a responsible life.

A year before his death in 2008, Edward Lorenz, the man who brought chaos theory to meteorology, was interviewed.

"I still don't hold much hope for day-to-day forecasting a month ahead," Lorenz said. "Two weeks ahead doesn't seem unreasonable at all now even though we haven't reached that point."

The statement was followed by a pause. "I gather," he added, "that a lot of the improvement has been from the improvement in initial conditions, and in turn, improvement in data assimilation methods."

Despite chaos theory, researchers keep trying. Some work on High-Resolution Rapid Refresh (HRRR) models. These models, among other things, pick up new radar data at fifteen-minute intervals with the goal of predicting small-scale hazardous weather—events such as tornadoes or violent storms carrying hail—that would be lost in the scale of lower-resolution models. Other researchers look at atmospheric waves miles above the ground, searching in particular for something they call a wave-number-5 pattern, which might be a multiweek harbinger of a severe heat wave. Still others watch for rain over the Indian Ocean and measure ocean temperatures, searching for a pattern called a Madden-Julian oscillation, which might—or might not—be a sign of future winter storms in California and Washington. Among the many programs supporting this work, there is the international collaboration called The Observing System Research and Predictability Experiment, or THORPEX, managed by the World Meteorological Organization with the goal of improving forecasts out to two weeks.

The quest continues.

At the far end of our little marina, I measure wind speed next to an old steel boat that sailed here from South Africa. The captain, an affable Dane, expresses curiosity. We chat. The conversation turns to forecasting and forecasts. He tells me that he never checks forecasts. He has, he says, no use for them. When it is time to go, he claims, he goes, prepared for anything. He believes his boat could withstand a hurricane.

Rust streaks stain the hull of his once handsome boat, and empty rum bottles line the cockpit. His boat's sails, rolled but not covered, are sun-faded. His inflatable dinghy floats behind his boat, its tubes soft. One of his dock lines, loose and dangling in the water, is green with thick algae. His boat, slowly becoming a hulk, has been bound to this dock, immobilized, for twenty-two months.

A Republican congressman, worried about damage and injuries and deaths from tornadoes in Oklahoma, worried about what he saw as the shortcomings of President Obama, lectured his colleagues in 2013. "Here's what we absolutely know," he said. "We know that Oklahoma will have tornadoes when the cold jet stream meets the warm Gulf air. And we also know that this president spends thirty times as much money on global warming research as he does on weather forecasting and warning. For this gross misallocation, the people of Oklahoma are ready to accept the president's apology."

The congressman might believe that the federal government is overfunding climate change research and underfunding forecasting research, but his numbers are off. In truth, the government probably spends something closer to three dollars on climate change research for every dollar spent on forecasting research. In the byzantine world of government cash flow, one cannot be too surprised by accounting errors, and in the Machiavellian theater of politics, one cannot be too surprised by hyperbole.

President Obama has yet to apologize to the people of Oklahoma.

Meanwhile, American meteorologists, the ones interested in next week's weather, also register complaints. Some, echoing the congressman from Oklahoma, complain about what they see as the imbalance between research funding for climate change and research funding for forecasting. The climate change researchers, they say, have better computer access. As they see it, climate change research gobbles up promising young scientists who could otherwise work on forecasting problems.

Despite scattered bitterness and bickering, climate change research and forecasting research share common ground. Climate change will, according to models confirmed by measurements, lead to a decrease in the temperature difference between the Arctic and the equator. The Arctic will warm more quickly than the equatorial and temperate zones. At a fundamental level, northern hemisphere winds are driven by the difference in temperature between the Arctic and the equator. A decrease in this difference, according to some researchers, will come with slower average wind speeds. But since temperatures will increase in the tropical and temperate zones, conditions that spawn hurricanes—high ocean temperatures—may become more common. Slower average winds could be accompanied by more hurricanes. Or not. Jet streams could be affected. Or not. Global average wind speeds could decrease while local average wind speeds in some areas could increase, or possibly vice versa.

Take, as a more fleshed-out example, another possible effect of an Arctic that is warming much faster than the temperate zone, an Arctic that is losing—has lost—much of its summertime sea ice. High above the ground, at a latitude just south of about sixty degrees north—the latitude of Anchorage, Alaska, and the southern tip of Greenland, and Oslo, Norway—a jet stream carries air, a lot of air, to the east. In that jet stream, in that river of air sometimes called the polar jet, winds can exceed one hundred miles per hour. When the temperature difference between the Arctic and the temperate zone decreases, the winds slow down. When the winds slow down, meanders in the polar jet increase. Those meanders are, of course, Rossby waves.

Since 1939, Rossby waves have been known to play a major role in weather over cities such as Boston, New York, Chicago, and Washington, DC. They are known to be important in forecasting. A Rossby wave that dips into the southeastern United States in winter can mean deep snow for East Coast urbanites. A Rossby wave that reaches far to the north over the west coast of North America can mean warm temperatures and light snow for the Pacific Northwest and Alaska. Both things happened in 2015.

Jeff Masters, the director of meteorology for Weather Underground, called it "jet stream weirdness" in an article he wrote for *Scientific American*. More important, he linked weather today and weather tomorrow to climate change. Weather and climate, though different, are inextricably intertwined. He also offered a fifteen-year forecast. He wrote, quite simply, "Expect the unprecedented."

But the concerns of American meteorologists are not all linked to climate change. In addition to issues with computers and data assimilation, as well as with the paucity of measurements at some locations, there are, some say, issues with leadership and government bureaucracies that ignore new findings. There are issues with funding sources that support basic research only until it begins to become useful, at which time new funding must be sought from sources that support routine forecasting, leaving promising projects stranded in what some meteorologists call "the Valley of Death."

Within the meteorological community, the grumbling and infighting of the past hundred years continues. There was the bitter battle between FitzRoy and his various detractors, the battle between James Espy and William Redfield, and the battle between Norwegian and American forecasters as D-day approached. Today, one battle focuses on climate change—not whether the climate is changing or whether the burning of fossil fuels contributes to that change, but whether climate change research should be supported at the expense of forecasting research. And there are other battles, small and large, over funding and pecking orders and nuanced technical details and unintended slights made over cocktails served at erudite conferences. Despite the beautiful complexity of their work, despite an entry gate that bars any but the smartest and hardest workers with the strongest academic backgrounds, the researchers are, after all, human beings.

But there are two important points to bear in mind before judging weather research and weather researchers. First, investments in forecasting pay high rates of return. Savings associated with forecasts come from things such as fewer unnecessary oil platform evacuations in the Gulf of Mexico, improved crop production, decreased injuries and loss of life during heat waves, decreased damage to or loss of ships and shipped goods during storms, and improved aviation efficiency. Estimated cost-benefit ratios for forecasting range from values as low as 2:1 — two dollars saved for every dollar spent — to 2,000:1. The World Bank Group estimates that improved forecasting could contribute up to thirty billion dollars per year in increased economic productivity and save an additional two billion dollars per year by reducing losses from storms. Internationally, forecasting research consumes something like two billion dollars each year, most of which goes toward satellites that serve more than one purpose. From an investment perspective, the math is far simpler than that normally confronted by forecasters.

Second, advances made so far cannot be ignored. A typical six-day forecast issued today is more accurate than a typical one-day forecast issued forty years ago.

Progress continues.

*Rocinante* is buttoned up. Her sails are down and bagged. Her portholes and hatches are dogged. Her bilge pumps are checked and rechecked. Soon, too soon, we will be away — first by bus, six hours to Guatemala City, and then by airplane, skipping quickly to the top of the troposphere, close to forty thousand feet, riding above that turbulent skin of troubled air that makes this planet's weather, headed back to our home in Alaska.

During that flight, our pilots will be informed by weather forecasts specifically tailored for aviation. They may talk of wind-optimized paths and wind-sculpted schedules. They may talk of ground miles and air miles, the first being the distance over ground from point to point, the second being the distance through moving air — shorter than the ground miles in a tailwind and longer in a headwind. At times, a flight-planning computer might issue a warning about inadequate fuel, but the fuel is only inadequate for ground miles, and a quick recalculation using air miles shows not only adequate but excess fuel. When the pilot estimates flight duration to the minute, he or she will rely not only on the plane's throttle and directions from air traffic controllers, but also on an understanding of constantly changing winds. In the United States alone, more than four thousand aviation weather forecasts go out each day. Complementing those forecasts, something like eight thousand graphics identify thunderstorms and areas where pilots should expect turbulence. At twenty-two Air Route Traffic Control Centers scattered across the nation, meteorologists work side by side with air traffic controllers to interpret threatening weather conditions as they appear and disappear. But that is not all. The pilots will also watch radar screens, piercing clouds to find storms, changing course here and there to avoid the worst of the bumps, which, after all, are nothing more than wind.

The information available on a routine flight would boggle the minds of those who once relied on telegraphs to generate simple weather maps. It would astonish the likes of Vilhelm Bjerknes and Lewis Fry Richardson. It would impress Jule Charney and John von Neumann and their colleagues who ran the first successful numerical forecast on ENIAC.

This final stage of our journey, traveling by bus and by airplane, seems like a simple thing. We will be taken care of by drivers and pilots. Our job descriptions will include words such as "sitting" and "eating." There will be no decisions about shortening sails, about interpreting weather forecasts, about safe distances from passing ships and lee shores. There will be no need for all-night vigils. There will be no place for an obsession with moving air. But this ending, this abandoning of our boat at a marina on a river in Guatemala, proves for me to be the most difficult part of the journey. These last few hours, with our checklists completed and our bags packed, seem worse than the worst parts of the journey so far, worse by far than a dragging anchor during a night of sideways pelleting rain.

On the dock, my co-captain comments on the same feeling. She talks of the sense of security that comes with home. She calls it "the illusion of security." Here on the boat, she says, a day could start with the promise of a champagne tour and end with a struggle, or vice versa. At home, plans tend to hold water, to materialize more or less on schedule. Here, they are interrupted by changing conditions, by unpredictable twists and turns. She says many things, but her words make it clear that she has not had enough of this sailing life. She left the dock in Texas thinking that life aboard would hold certain attractions, and now, leaving the dock in Guatemala, she knows that she was right. It is a poignant moment for us, two co-captains perfectly synchronized in our view of the world in front of a boat that we will leave behind for a time.

Leaving Galveston, we hoped to live closer to the elements and to learn something about the wind. We have done both. We have learned valuable practical lessons, too, making us better sailors. And we have relearned something else, something less practical but more valuable. We have relearned that the close examination of things we accept as part of everyday life exposes beautiful complexities. Careful scrutiny of the background noise that permeates our daily existence reveals exquisite intricacies. In examining wind, we have sampled layers of knowledge that build on one another, seemingly without end. In exploring forecasts, we have glimpsed a history that continues to unfold. We have relearned child-like wonder.

Back under the palapa, I watch a thunderstorm approach. A few fat precursory drops of rain fall, plopping into the shallow water in front of the palapa, radiating ripples through the sparse stand of bulrushes that grow between my perch and the dock. Those drops have been on a journey of their own, forming in the sky, blowing sideways and upward and sideways again, eventually succumbing to gravity, abandoning the winds in exchange for the comfort of the wide lake-like river.

I watch as the wind increases its pace. It does so in bursts, here and there, expressed as sudden cat's-paws on the water's surface, as gusts that twist the leaves of trees or that disturb the bulrushes, bending their tips for a second or two before diminishing and letting the spikes spring back to attention. The gusts express turbulent flow in the atmosphere, the unevenness of air, patches of low pressure suddenly formed and filled.

Neither the mainstream of air moving toward me nor the minor streams moving up and down and sideways and backward flow coherently. The streams are groups of individual molecules, each molecule stubbornly independent, each finding its own way, running in the middle of the main pack now, then looping upward or sideways, perhaps staying, overall, with the pack, or perhaps riding an eddy for a while to join another pack, another parcel of air. On average, though, the molecules move toward me, and any fool would say the wind is from the northeast and the air is moving southwest, because we experience life in generalities, in a series of averages that leave manageable impressions. We experience wind as coherent streams, as if the molecules were attracted to one another, one following the next like obedient sheep, but that is not at all what happens. Air molecules are not attracted to other air molecules. And while air tends to move from higher pressure toward lower pressure, to fill voids, the movement is turbulent, especially near the ground and across the water's surface, not regular at all, not as predictable as one might hope at scales of time and space important to daily experience, to sailing, to life.

In the comfort of the palapa, on a swinging chair that hangs from a beam, I watch as the storm approaches and its fat drops fall. While watching, I reread passages from Daniel Defoe's *The Storm*. Late in 1703, Defoe advertised to find people who had experienced the recent storm, the recent wind. He tried to understand it. He commented beautifully and accurately on human knowledge of moving air. There in the palapa, I reread a particular passage, a favorite string of Defoe's words. "Those Ancient Men of Genius who rifled Nature by the Torch-Light of Reason even to her very Nudities," he wrote, "have been run aground in this unknown Channel; the Wind has blown out the Candle of Reason, and left them all in the Dark."

Since Defoe, astounding progress has been made. The candle of reason may falter now and then in the breeze of bungled forecasts and failed satellites and research dead ends, but its flame glows.

I look out upon the water, thinking of forecasters and researchers, of people who think about moving air. I think of the thousands who work the weather now, the Chris Parkers, the unnamed forecasters at PassageWeather and Weather Underground. I think of sailmakers and sailors, of wind turbine installers and Flettner rotor designers. I think about Lorenz and Charney, Bjerknes and Rossby, FitzRoy and Ferrel. I think of Richardson near the front lines in World War I, scribbling numbers, imagining his grand theater of forecasters.

I watch the fat drops of rain become a downpour. A gust of wind hits the palapa, carrying water with it, spattering the pages of my book. Closing Defoe, I abandon my chair and step backward, farther into the protection of the palapa, and from there I watch the storm move over the Río Dulce.

# ACKNOWLEDGMENTS

My personal voyage to discover wind did not occur in a vacuum. In fact, it would not have been possible without the assistance and encouragement of quite a crowd.

My late father, perhaps without realizing it, passed along a love of the water and a desire to sail across oceans. From start to finish, I would have been entirely lost without my wife and co-captain, Lisanne Aerts. And while the likes of Tony Smythe, Kevin Churchill, and a host of others in the sailing community may be entirely unaware of their importance in the course of our lives, without them Lisanne and I may never have bought *Rocinante* and set out upon our small journey.

All authors are forever indebted to their editors, and I am no exception. John Parsley, my editor at Little, Brown, has not only provided valuable guidance, but he has also had the courage to believe in and encourage an unknown author with an unusual (or so it seems to me) style. I consider our association to be one of the highlights of my professional life. In addition, I would like to express my appreciation of Keith Hayes, who designed the cover of *And Soon I Heard a Roaring Wind*—a cover that brought a smile to my face the first time I saw it and the hundredth time I saw it. I am, of course, also indebted to the entire team at Little, Brown, including Elizabeth Garriga, Betsy Uhrig,

Malin von Euler-Hogan, and others who work behind the scenes to make authors like me look far better than we deserve.

A number of people commented on sections of an early draft of the manuscript that became *And Soon I Heard a Roaring Wind*. Commenters included many of those mentioned in the paragraphs above, but also Lucy Slevin and Kath Temple, along with fellow writers Deb Vanesse and Don Rearden, my good friend and fellow sailor Jason Hale, and my good friend and diving partner Bill Lee. My son, Ish Streever, although he does not appear in the book, was aboard *Rocinante* for a short leg of the voyage along the coast of Mexico, but more important, he encouraged me when my own doubts about my writing slowed me down, and he provided insightful comments on an early draft of the manuscript.

Barbara Jatkola provided detailed line edits and fact checks on a later draft of the manuscript. I cannot overstate the value I place on her attention to detail, but to stop there would do justice to neither her eye for broader issues that warranted attention nor to her patience with boneheaded mistakes.

For my understanding of wind, I am indebted to dozens of currently active scientists and forecasters, including, among many others, Chris Parker, Peter Lynch, Jeff Masters, Cliff Mass, Robert Fovell, Lee Chesneau, the entire staff of Passage-Weather, and the entire staff of Weather Underground. I would be remiss if I did not extend this acknowledgment to the entire community of meteorologists. In my experience as a scientist, it seems to me that this community, more so than any other community of scientists, has for a long time pursued and continues to pursue outreach programs intended to inform the public about a very difficult and dramatically underappreciated field.

The illustrations accompanying the text come from many sources. Wikimedia Commons and the Library of Congress deserve special notice for compiling valuable historical photographs and drawings, many of which are available for use by authors like me. I also thank Professor Lennart Bengtsson for allowing the use of his drawing of Richardson's imaginary forecasting theater, Derek Ogle and Paul Brooker for allowing the use of the photograph of the *Weather Adviser,* and the Kelvin Smith Library for allowing the use of the Brush windmill photograph.

I sometimes wonder what it would be like to have everyone who has influenced *And Soon I Heard a Roaring Wind* in one room, accompanied by the likes of some of the long-dead characters who populate the book, including historical figures such as Halley, Hadley, FitzRoy, Defoe, Slocum, Richardson, Darwin, Lorenz, Charney, Caldwell, Ferrel, Espy, Redfield, Columbus, Dickens, von Neumann, Beaufort, Brush, Galton, Morse, Arago, Le Verrier, Voss, Rossby, Krick, Petterssen, Glaisher, Swan, Galileo, Hedin, Bjerknes (both father and son), and many others. Like Lewis Fry Richardson dreaming of his forecasting theater, I dream of a gathering that would require a rather large room. No doubt my dream, if it were to become reality, would be a strange but memorable party.

# NOTES

## Introduction: Before Departure

John J. Miller, a journalism professor, wrote about Daniel Defoe's *The Storm; or, A Collection of the Most Remarkable Casualties and Disasters Which Happen'd in the Late Dreadful Tempest, Both by Sea and Land* in the August 13, 2011, issue of the *Wall Street Journal*, just after Tropical Storm Emily died over Cuba and at a time when forecasters predicted months of high hurricane activity. "Defoe's eyewitness account," Miller wrote, "is valuable, but his real innovation was to collect the observations of others." He also noted that *The Storm*—a book that remains available today and that is often cited by students of weather and climate—did not sell well when it was first published. As he put it, all books are "vulnerable to the whims of consumers," a reality that every writer understands. Defoe, just before writing *The Storm*, had been pilloried and then imprisoned at Newgate for the crime of writing a pamphlet considered to be seditious, so the poor performance of his book must have been especially depressing.

In the preface to his 1922 book *Weather Prediction by Numerical Process* (Cambridge University Press, Cambridge), Lewis Fry Richardson compared predictions of astronomical motion with predictions of weather. "It would be safe to say," he wrote, "that a particular disposition of stars, planets and satellites never occurs twice. Why then should we expect a present weather map to be exactly represented in a catalog of past weather?" Instead, he turned to mathematics.

David McCullough's excellent 2015 book *The Wright Brothers* (Simon & Schuster, New York) emphasizes the importance of wind to early fliers. Too little wind, and flying was challenging. Too much wind, and it was impossible. The invention of human flight (and possibly the evolution of flight in other organisms) required Goldilocks wind conditions.

## Chapter 1: The Voyage

Erik Larson's wonderful *Isaac's Storm: A Man, a Time, and the Deadliest Hurricane in History* offers a detailed account of the 1900 hurricane that destroyed

Galveston. Other valuable book-length accounts of the storm include *Story of the 1900 Galveston Hurricane,* edited by Nathan C. Green (1900; republished in 2000 by Pelican, Gretna, LA); John Edward Weems's *A Weekend in September* (1957, Holt, New York); and *Through a Night of Horrors: Voices from the 1900 Galveston Storm* (2000, Rosenberg Library, Galveston), a collection of firsthand accounts edited by Casey Edward Greene and Shelly Henley Kelly.

The September 9, 1900, *New York Times* coverage of the storm that destroyed Galveston was headlined "Great Disaster at Galveston. Deaths May Be over 2,600—4,000 Houses Ruined. A Heavy Property Loss." This story, published one day after the storm, shows that news traveled very quickly even in 1900.

The name "Rocinante," like so many aspects of the novel *Don Quixote,* has layered meanings. *Rocin* is Spanish for what many Americans would call a nag—a worn-out workhorse—but according to many sources, it can also refer to a person with the characteristics of a worn-out horse. *Ante* can, of course, mean "before." In the novel, Cervantes himself explains his intent in calling the horse Rocinante, writing that it was "a name, to [Don Quixote's] thinking, lofty, sonorous, and significant of [the horse's] condition as a hack before he became what he now was, the first and foremost of all the hacks in the world."

Raymond Chandler's quote about wind is from the opening of his short story "Red Wind," first published in the January 1938 issue of *Dime Detective Magazine.*

Derechos have winds as powerful as those of some hurricanes and tornadoes. "Derecho" comes from the Spanish word meaning "straight."

Defining "sustained wind speeds" is more complicated than one might expect. A World Meteorological Organization paper called "Definition of Maximum Sustained Wind Speed of Tropical Cyclones" (presented at the Sixth Tropical Cyclone RSMCs/TCWCs Technical Coordination Meeting, Brisbane, Australia, November 2–5, 2009) will enlighten tolerant readers. It is available online at http://www.wmo.int/pages/prog/www/tcp/documents/Doc2.3_WindAveraging.pdf.

Captain John Smith's words are from his 1627 *A Sea Grammar,* republished in 1691 as *The Sea-Man's Grammar and Dictionary, Explaining All the Difficult Terms in Navigation: and the Practical Navigator and Gunner: In Two Parts.* The book was a technical glossary for sailors. The 1691 edition, published by Randal Taylor of London, is available online at http://www.shipbrook.net/jeff/bookshelf/details.html?bookid=27. It contains many words that remain familiar today, such as "hawser" and "furling lines," but also words that are less familiar, such as "loofehook" ("a tackle with two hooks, one to hitch into a chingle of the main or fore sail, in the bolt rope in the leech of the sail by the clew, and the other to strap spliced to the chefters to boufe or pull down the sail").

Most of my knowledge about the detailed history of the Beaufort scale began with Scott Huler's wonderful book *Defining the Wind* (2004, Three Rivers Press, New York). To Huler, the scale is more than just a scale. It is "a philosophy of attentiveness, a religion based on observation: an entire ethos in 110 words. One hundred ten words, that is, and four centuries of backstory." He saw poetry in what is usually presented as Beaufort's original version of the scale, but he realized that the poetry predated Beaufort. Beaufort, seeking a scale that would make sense to the average British naval officer, took the best of what his many predecessors offered. Huler also succinctly describes modern weather awareness: "Someone asks what the weather will be, and instead of walking outdoors, we log on to weather.com." Another valuable Beaufort scale source is F. Singleton's "The Beaufort Scale of Winds—Its Relevance, and Its Use by Sailors" (2008, *Weather*, vol. 63, no. 2).

Tycho Brahe lost his nose in a duel on December 29, 1566. The duel was intended to resolve an argument over a mathematical formula. As a consequence, he wore a metal nose (possibly of copper, silver, or gold) pasted to his face. The possibility that he inspired *Hamlet* is usually tied to his death and his assumed or alleged affair with a Danish queen, but at least one author has pointed out interesting astronomical allusions in *Hamlet* that could be tied to Brahe. Peter D. Usher, a professor from Pennsylvania State University, suggested that *Hamlet* is at least partly an allegory of competition between two cosmological models, one of which was Brahe's.

Captain John Voss wrote about his various adventures in *The Venturesome Voyages of Captain Voss* (1913, Dodd, Mead, New York). His dugout canoe, named the *Tilikum*, is now displayed at the Maritime Museum of British Columbia in Victoria.

Larry Ellison's words about the value of winning and losing the America's Cup race are from the documentary *The Wind Gods: 33rd America's Cup* (2013, Skydance Productions). Skydance Productions is owned by Larry Ellison's son, David. Skydance is also behind films such as *Star Trek into Darkness* (2013) and the 2010 remake of *True Grit*.

Sir Francis Beaufort's instructions are from Captain Robert FitzRoy's *Narrative of the Surveying Voyages of His Majesty's Ships* Adventure *and* Beagle *Between the Years 1826 and 1836, Describing Their Examination of the Southern Shores of South America, and the* Beagle's *Circumnavigation of the Globe*, vol. 2: *Proceedings of the Second Expedition, 1831–1836, Under the Command of Captain Robert Fitz-Roy, R.N.* The book was originally published in 1839 (Henry Colburn, London) as a multivolume set that included Darwin's account of the trip, which was later sold on its own under various titles, including *The Voyage of the* Beagle. FitzRoy's book remains available online at http://darwin-online.org.uk/converted/published/1839_Voyage_F10.2/1839_Voyage_F10.2.html and perhaps elsewhere.

Beaufort, when asked to find a scientist to sail with the *Beagle,* turned to George Peacock at Cambridge. Peacock contacted the Reverend John Stevens Henslow, who knew Darwin from dinner parties and from lectures on mathematics, theology, and botany. Peacock wrote to Darwin in 1831, before the voyage of the *Beagle:* "I received Henslow's letter last night too late to forward it to you by the post, a circumstance which I do not regret, as it has given me an opportunity of seeing Captain Beaufort at the Admiralty (the Hydrographer) & of stating to him the offer which I have to make to you: he entirely approves of it & you may consider the situation as at your absolute disposal: I trust that you will accept it as it is an opportunity which should not be lost & I look forward with great interest to the benefit which our collections of natural history may receive from your labours." The letter goes on to outline the planned voyage of the *Beagle* to Tierra del Fuego and the South Seas. "The expedition," wrote Peacock, "is entirely for scientific purposes & the ship will generally wait your leisure for researches in natural history." He also noted that it was an unpaid position: "The Admiralty are not disposed to give a salary, though they will furnish you with an official appointment & every accommodation." Then as now, biologists were underpaid.

## Chapter 2: The Forecast

Edmond Halley's description and map of trade winds appeared in a 1686 issue of the *Philosophical Transactions* of the Royal Society of London under the title "An Historical Account of the Trade Winds, and Monsoons, Observable in the Seas Between and Near the Tropicks, with an Attempt to Assign the Phisical Cause of the Said Winds."

Benjamin Franklin's ideas about winds and storms are from his letter to the Reverend Jared Eliot dated July 16, 1747. The upcoming release of the Evans map was advertised in Franklin's newspaper, the *Pennsylvania Gazette,* on October 13, 1748: "The Map of Pennsilvania, New Jersey and New York Provinces, by Mr. Lewis Evans, is now engraving here." The first edition of the map contained Franklin's one-sentence description of his ideas about winds and storms, but subsequent editions, which required space for increased topographical detail, dropped the description. Franklin did not claim credit for the words on the map. In 1905, William Morris Davis weighed in on what must have been a wider discussion regarding credit for the words. From Davis's talk, published in the *Proceedings of the American Philosophical Society* in May 1906 and titled "Was Lewis Evans or Benjamin Franklin the First to Recognize That Our Northeast Storms Come From the Southwest?": "It would thus appear that Franklin contributed a statement of his discovery to Evans' map, making no claim whatever for recognition or priority; and indeed, suffering the statement to be obliterated without remonstrance, so far as now appears, when the second edition of the map was published. Generous as he thus showed himself, to the point of indifference, it is still fitting that we at this time should take pains to give credit where credit is due." Another important source about Franklin's life in general is J. A. Leo Lemay's *The Life of Benjamin Franklin,* vol. 2: *Printer and Publisher, 1730–1747* (2006, University of Pennsylvania Press, Philadelphia).

The description of the wreckage from the storm of 1854 in the Black Sea comes from Edward Henry Nolan's 1857 book *The Illustrated History of the War Against Russia* (James S. Virtue, London; reprinted on paper and electronically and still readily available). The book, running eight hundred pages, contains beautiful illustrations, ranging from portraits to battle scenes. Its words capture the horror of the war made famous by Alfred, Lord Tennyson, in his "Charge of the Light Brigade," which describes a battle that occurred before the storm hit.

The 1854 ships' log entries are from "Great Historical Events That Were Significantly Affected by the Weather: 5, Some Meteorological Events of the Crimean War and Their Consequences," by S. Lindgrén and J. Neumann (1980, *Bulletin of the American Meteorological Society*, vol. 61, no. 12).

Samuel Morse's letter to his brother, who was in England at the time, was written on April 19, 1848. The letter includes mention of Morse's plans to litigate: "A trial in court is the only event now which will put public opinion right, so indefatigable have these unprincipled men been in manufacturing a spurious public opinion." Despite all this, Morse was already accumulating wealth as the result of his invention. The letter also describes a house he purchased, "a home, a beautiful home, for me and mine," unencumbered by a mortgage. Morse was an artist and a Calvinist as well as an inventor. This is not to say he was open-minded or progressive. He spoke in favor of slavery, and he spoke against Catholic immigration to the United States. His letters were collected and published in 1914 under the title *Samuel F. B. Morse, His Letters and Journals in Two Volumes,* edited and supplemented by his son Edward Lind Morse. Portions of Morse's letters can be viewed at http://www.fulltextarchive.com/page/Samuel-F-B-Morse-His-Letters-and-Journalsx5161/.

François Arago's words dismissing the possibility of weather forecasting are from his 1846 article in the *Edinburgh New Philosophical Journal* (vol. 41, pp. 1–16), wonderfully titled (because of the title's length, and because it directly poses two very different questions, one practical and one philosophical) "Is It Possible, in the Present State of Our Knowledge, to Foretell What Weather It Will Be at a Given Time and Place? Have We Reason, at All Events, to Expect That This Problem Will One Day Be Solved?" He wrote the article in part to defend himself against published weather and climate forecasts illegitimately attributed to his name: "I must frankly own, that I wished to have an opportunity of protesting decidedly against the predictions which have every year been attributed to me, both in France and in other countries." His discussion covers topics as diverse as sea ice, temperature variations in Paris, "animals of the medusa kind" (presumably jellyfish), and the effects of wind on fruit trees on the Isle of France. He also speculates on human-induced climate change: "I have considered whether the operations of man, and occurrences which will always remain beyond the range of our foresight, might not be of such a nature as to modify climates accidentally, and in a very sensible manner, in particular with regard to temperature. I already perceive that facts will answer in the affirmative." The article is readily available online.

The story of Le Verrier's weather forecasting has been summarized by many authors. Two books with good summaries are *Invisible in the Storm: The Role of Mathematics in Understanding Weather*, by Ian Roulstone and John Norbury (2013, Princeton University Press, Princeton, NJ), and *Storm Watchers: The Turbulent History of Weather Prediction from Franklin's Kite to El Niño*, by John D. Cox (2002, John Wiley and Sons, Hoboken, NJ). Although not every detail about Le Verrier is consistent across the many retellings of his story, most authors paint him as both arrogant and insecure, a deadly combination in anyone, but especially harmful in someone entrusted with managing a scientific institute such as the Paris Observatory.

The passage describing FitzRoy's view of risk comes from his aforementioned 1839 *Narrative of the Surveying Voyages of His Majesty's Ships* Adventure *and* Beagle *Between the Years 1826 and 1836*, vol. 2, http://darwin-online.org.uk/converted/ published/1839_Voyage_F10.2/1839_Voyage_F10.2.html. These words offer some understanding of FitzRoy's willingness to take on the challenge of forecasting, both in the sense of providing a service to captains at sea and in the sense of hinting at his perception of a personal space in history. Peter Moore's 2015 book *The Weather Experiment* (Farrar, Straus and Giroux, New York) brought the passage to my attention. Moore provides an admirably detailed account of many of the characters mentioned in this book, including Beaufort and FitzRoy.

FitzRoy's letter to Darwin regarding the inclusion of advertisements in their four-volume book was written on March 20, 1839. He began the letter by objecting to the publisher's inclusion of advertising but was quick to say that he supported the inclusion of advertisements for other books that Darwin himself was preparing. "I certainly objected as strenuously as I could to Mr. Colburn's loading our already too thick volumes with a pack of advertisements—such as one sees in every Review or monthly Magazine—but there is surely every reason for my wishing that an advertisement of the Zoology which you are bringing out—as well as of your forthcoming Work on Geology—should be attached to some part of the three volumes." This letter, as well as others, including letters to and from the *Beagle*, is available through the Darwin Correspondence Project, http:// www.darwinproject.ac.uk/explore-the-letters.

John William Draper presented "The Intellectual Development of Europe, Considered with Reference to the Views of Mr. Darwin and Others, That the Progression of Organisms Is Determined by Law," which set the stage for the famous exchange between Wilberforce and Huxley in 1860. There are many accounts of that exchange, and there can be no doubt that it is a story that grew with the telling. Even today, scholars write of what was and what was not actually said, and of its significance.

Saxby's *Foretelling Weather: Being a Description of a Newly-Discovered Lunar Weather-System* (Longman, Green, Longman, and Roberts, London) was first published in 1862. Although he is seen today as little more than an astrologer, Saxby was taken very seriously by both himself and his supporters in the 1860s. "Hitherto the public thirst to possess a knowledge of coming weather has tempted many an impostor to vend his spurious prognostics," he wrote on page 5 of his book, "the constant failure of which still further prejudice the whole question; hence the difficulty of obtaining a hearing for any suggestions founded on really honest deductions." The second edition of this book (published in 1864) was edited in part to defend his ideas against criticism by FitzRoy and others.

Francis Galton's 1874 book is *English Men of Science: Their Nature and Nurture* (Macmillan, London). Galton's ideas about eugenics were later promoted by Charles Davenport, whose work led to the sterilization of humans deemed undesirable to the gene pool. They also contributed to the horrific Nazi application of the principles of what was called "racial hygiene" and ultimately the Holocaust. Later in life, Galton drafted an unpublished novel about a Utopian society based on selective breeding of humans. His niece burned most of the manuscript.

Heliostats are devices used to direct the sun's rays toward a distant target. Today they are used in some solar energy applications. Galton described his heliostat in an 1858 talk, which was followed by an article titled "A Hand Heliostat, for the Purpose of Flashing Sun Signals from On Board Ship, or on Land, in Sunny Climates," published in *The Engineer,* a publication of the British Association for the Advancement of Science. The following year, he published an article describing an improved version of the same device. Even to a novice sailor, it appears to be a gadget that would be difficult to use at sea and that would rarely offer advantages over standard methods of signaling other ships with flags. I have not come across any references suggesting that Galton's device was ever used at sea.

James Glaisher and his colleagues started the Daily Weather Map Company Limited in 1861 or earlier. According to the October 1901 issue of the *Quarterly Journal of the Royal Meteorological Society* (vol. 27, no. 120), in what appears to be a description of exhibits displayed during a meeting of the society, "This Company was formed for the purpose of raising capital to carry on the publication of the *Daily Weather Map.* The capital was to be £4000 in 400 shares of £10 each. Two maps were printed, viz, those for August 5 and September 3, 1861."

FitzRoy's forecasts were also attacked by men in the salvage business, who feared that accurate forecasts would save ships and ultimately drive them out of business.

Criticism of FitzRoy's forecasts has been widely reported by biographers, but actual examples are hard to find. The *Times* article of June 18, 1864, came to my attention via John and Mary Gribbin's biography, *FitzRoy: The Remarkable Story of Darwin's Captain and the Invention of the Weather Forecast* (2004, Headline

Books, London). When I found this book, I was delighted to see that I had read and enjoyed two other books by the same authors, *Richard Feynman: A Life in Science* and (by John Gribbin, working on his own) *In Search of Schrödinger's Cat,* a wonderful explanation of quantum physics, which, had I been born with a more efficient brain, might have led me down an entirely different path in life.

Darwin remarked on FitzRoy's mental health long before his suicide. In the postscript of a letter from Darwin to the renowned geologist Charles Lyell dated August 9, 1838, he wrote of FitzRoy: "I never cease wondering at his character, so full of good & generous traits but spoiled by such an unlucky temper. Some part of the organization of his brain wants mending: nothing else will account for his manner of viewing things."

## Chapter 3: Theorists

All of the Edmond Halley quotes are from his aforementioned 1686 article, "An Historical Account of the Trade Winds, and Monsoons, Observable in the Seas Between and Near the Tropicks, with an Attempt to Assign the Phisical Cause of the Said Winds." Halley, in writing about the trade winds, felt that a causal explanation was obligatory: "But least I should seem to propose to others, difficulties which I have not thought worth my own time and Paines, take here the result of an earnest endeavour after the true reason of the aforesaid *Phenomena.*" And he knew he might be wrong: "If I am not able to account for all particulars, yet 'tis hoped the thoughts I have spent thereon, will not be judged wholly lost, by the curious in Natural Inquiries."

George Hadley's words were published in the 1735 article "Concerning the Cause of the General Trade-Winds," published in the *Philosophical Transactions* of the Royal Society of London (vol. 39, pp. 58–62). He begins the paper by recognizing previous work, presumably including that of Edmond Halley. "I think the Causes of the General Trade-Winds have not been fully explained by any of those who have wrote on that Subject, for want of more particularly and distinctly considering the Share the diurnal Motion of the Earth has in the Production of them: For although this has been mention'd by some amongst the Causes of those Winds, yet they have not proceeded to shew how it contributes to their Production; or else have applied it to the Explication of these Phenomena, upon such Principles as will appear upon Examination not to be sufficient." In other words, his predecessors had a hunch about the role of a rotating earth in generating the trade winds, but little more.

Espy believed he could control storms through the judicious use of forest fires. He proposed the setting of fires to initiate rainstorms. Through his lecture tours, he became known in the public mind as "the Storm King" and "the Old Storm King." The description of Espy comes from his friend Alexander Dallas Bache, according to the aforementioned *Storm Watchers: The Turbulent History of Weather Prediction from Franklin's Kite to El Niño.*

William C. Redfield's article about the storm was published in 1831 in the *American Journal of Science and Arts* under the title "Remarks on the Prevailing Storms of the Atlantic Coast of the North American States." Earlier reports of the circular nature of hurricanes had been offered by ship captains and others, but Redfield's report drew the attention of the scientific world. Benjamin Silliman, who would play a key role in the development of the world's first oil well, was the editor of the journal. The issue also included an article about steamboat safety, an important topic in an era of exploding boats and probably of special interest to Redfield, who was associated with steam-powered shipping. Redfield's 1839 letter was published in the *Journal of the Franklin Institute* (vol. 27, no. 6, pp. 363–78) and titled "Remarks on Mr. Espy's Theory of Centripetal Storms, Including a Refutation of His Positions Relative to the Storm of September 3rd, 1821: With Some Notice of the Fallacies Which Appear in His Examinations of Other Storms." Redfield did not use a middle initial until later in life, when he adopted the "C" to differentiate himself from other William Redfields. According to at least one source, he claimed that it stood for "Convenience," although it is usually reported to stand for Charles.

Joseph Henry wrote about the dispute between Espy and Redfield in *Results of Meteorological Observations, Made Under the Direction of the United States Patent Office and the Smithsonian Institution, from the Year 1854 to 1859, Inclusive, Being a Report of the Commissioner of Patents Made at the First Session of the Thirty-Sixth Congress*, vol. 2, part 1 (1861, Government Printing Office, Washington, DC). Espy and Redfield were both dead before the report was published.

The words regarding the reconciliation of the ideas of Espy and Redfield are from "Article VII of the Report of the Tenth Meeting of the British Association for the Advancement of Science, Held at Glasgow, Sept. 1840," which appeared in a publication of the British Scientific Association. These words are sometimes attributed to Joseph Henry.

The biographical information on Ferrel is from *Memoir of William Ferrel, 1817–1891*. The memoir was presented by the famous meteorologist Cleveland Abbe to the National Academy in April 1892, soon after Ferrel's death, but much of it was autobiographical, written by Ferrel in 1888 at the request of a friend. Ferrel's brother's words are also from the memoir, which is available online at http://www.nasonline.org/publications/biographical-memoirs/memoir-pdfs/ferrel-william.pdf. Abbe links Ferrel's genius to evolution: "What a profound problem in psychology is suggested to us by the reflection that, in all ages of the world, so many persons are born who instinctively give themselves to searching out nature's laws and methods. The students of evolution have endeavored to demonstrate some few principles in regard to the inheritance of scientific ability, but who shall explain, in a direct and acceptable manner, the remarkable outcrop, here and there, of individuals whose ancestry gives no evidence of the possession of those qualities of mind that distinguish the brilliant descendant. The botanist singles out a peculiar flower or head of grain and calls it a sport, an exceptional individual among ten thousand of its comrades." In other words, as far as Abbe was concerned, Ferrel was a sport.

The report commissioned by Joseph Henry was written by Elias Loomis in 1847. It was called, simply enough, "Report on the Meteorology of the United States," and it was presented as appendix 2 in the *First Report of the Secretary of the Smithsonian Institution to the Board of Regents; Giving a Programme of Organization, and an Account of the Operations During the Year.*

Sir Napier Shaw, a British meteorologist, is credited with coining the term "geostrophic," meaning "spinning earth." He also introduced the millibar as a convenient unit of pressure. He wrote, with John Switzer Owens, *The Smoke Problem of Great Cities* (1925, Constable, London), which, while not the first volume written about air pollution, was an important contribution. The flyleaf features an image of a man burdened by a heavy burlap sack (filled with soot?), with the caption, "The soot-fall in a minute within the county of London and a Londoner on the same scale." The paper version of the book is available from various suppliers. An electronic version is available online at https://archive.org/stream/smokeproblemofgr00shaw#page/n19/mode/2up.

## Chapter 4: Initial Conditions

Regarding Galileo and the weight of air, from a letter to the editor published in *Nature* (1908, vol. 78 [July 30], p. 294): "The discovery, in the first half of the seventeenth century, that the air has weight is associated with things of immense importance, for instance, the invention of the barometer and the refutation of the dogma—dear to the false science and the false philosophy of the day—that 'Nature abhors a vacuum.' In a new edition of the 'Essais de Jean Rey,' reviewed in NATURE of July 9, an attempt is made to assign this discovery to Rey, and, so far to regard Torricelli, Galileo, Pascal, and Descartes as his disciples. Without claiming to be an authority upon Rey or upon Galileo, I would direct attention to the statement, made in 'Galileo—His Life and Work,' by J. J. Fahie, that Galileo's way of determining the specific gravity of the air was first described in his letter to Baliani dated March 12, 1613. Rey's 'Essais' was published in the year 1630."

My much-abbreviated history of the barometer, which leaves out many steps and many nuances, comes from many sources. Among these is a very enlightening and well-illustrated article published in 1944, "A Brief History of the Barometer," by W. E. Knowles Middleton, published in the *Journal of the Royal Astronomical Society of Canada* (vol. 38, no. 2 [February], pp. 40–64). It is available online at http://adsabs.harvard.edu/full/1944JRASC..38...41K. The journal, now published every other month, is more than a century old. As far as I know, it no longer considers papers on meteorology.

There are about sixty-four species of flying fishes, depending on exactly how one divides species and what one considers to be a true flying fish. All are in the family Exocoetidae.

The article in *Science* correcting Robinson was written by C. F. Marvin (1889, vol. 13, no. 231, p. 248). Other titles in the same issue include "The Velocity of Storms as Related to the Velocity of the General Atmospheric Movements" (a letter, which concluded with the statement, "I trust these few facts may serve to further stimulate the interest which is now being aroused in more exact and detailed cloud observations") and "A Meteorological Exhibition" (describing the New England Meteorological Society's fourteenth regular meeting).

Most of the information on weather ships is from "History of the British Ocean Weather Ships," by Captain C. R. Downes, published in 1977 by the *Marine Observer* (vol. 47, pp. 179–86). Downes provides specific details about each weather ship, including the original name of the ship, its new name as a weather ship, and its size, sailing dates, and assigned locations. There are photographs of two of the ships. As is often true of accounts written by professional mariners, his includes few personal details about life on board these ships, possibly because such a life can seem mind-bluntingly routine to those who actually live it. Another source, which does offer glimpses into the life of the weather ship crews, is the October 1949 issue of *Weather Bureau Topics* (vol. 8, no. 46), a publication of the US Weather Bureau (available online at http://docs.lib.noaa.gov/rescue/wb_topicsandpersonnel/1949.pdf). This publication in general—at least in 1949—seems to have been remarkably focused on retirements. However, the October issue includes a short article called "Ocean Weather Duty," which gives the ranks of the meteorological staff aboard the vessels and notes that weather patrols normally carried "a total of five Weather Bureau men." An assignment to a weather vessel typically ran for one year, but "preferably two or more years." The observers were based in Boston, New York, Norfolk, and San Francisco, although they sometimes traveled to and from other ports where they joined and left the ships. Ships remained on station for twenty-one days, so each trip, including steaming to and from the station, ran from twenty-seven to about thirty-seven days. Crew members worked sixty-three hours each week, with twenty-three of those hours paid at the Weather Bureau's authorized overtime rate. They were given a per diem of $2.40, an amount deemed "adequate to pay the cost of meals, laundry, and other incidentals" aboard ship. Their status aboard seems to have been equivalent to that of wardroom officers—that is, one notch above midshipmen, who were the junior-most officers (sometimes cadets) aboard. Ashore, between missions, the men were assigned other duties with the Weather Bureau. "Duty with these patrols," according to the unnamed author of the article, "has also proved an interesting assignment for weathermen with a taste for sea life." A third source is the excellent website built by the son of Fredric Brooker, who worked as a radio technician on a weather ship. The site, Ocean Weather Ships (http://www.weatherships.co.uk/), includes dozens of photographs of weather ships and tracks the refurbishment and renaming of ships. One ship, which started life as the *Amberley Castle* and then became the *Weather Adviser,* was ultimately renamed the *Admiral FitzRoy.* The *Admiral FitzRoy* was scrapped in 1982. The quotations about northern lights, waves coming over the bridge, and walking back and forth are from this website and are attributed to Andy Reilly, who was stationed on the *Surveyor* in the mid-1960s. Mr. Reilly, apparently late in life when he provided information for the website, ended his memories by saying, "I wished I had stayed." That is, he regretted leaving the weather ship service.

Anton Eliassen was the director of the Norwegian Meteorological Institute who commented on the absence of offers to assist with funding of the *Polarfront.* His comments, and those of scientists lamenting the cessation of the *Polarfront's* mission, were reported by Quirin Schiermeier in *Nature News,* June 9, 2009.

Accounts of James Glaisher's ballooning adventures (or antics, depending on one's viewpoint) are captured in the 1871 book *Travels in the Air,* written by Glaisher, Camille Flammarion, W. de Fonvielle, and Gaston Tissandier (Richard Bentley & Son, London). Both paper and electronic versions of the book remain available, including the electronic version of the first edition at https://ia802708.us.archive .org/9/items/travelsinair00glaigoog/travelsinair00glaigoog.pdf. The description of the windy ascent in the *Captive* comes from a portion of the book attributed to de Fonvielle and Tissandier. The first edition of the book, which includes 118 illustrations, is a mix of nearly poetic aerial travelogue and remarkably tedious descriptions of data. For example, it includes beautiful cloud and clear sky imagery, but also an entire page describing travel speeds between destinations, which could have been presented in a single short table. Throughout, Glaisher (who edited the entire volume) appears heroic. Like many men of history, he was his own best publicist.

In 1939, a Weather Bureau employee named George Mindling, in a collection of weather poems, wrote about early automated weather balloons. The first four lines compare balloons and airplanes:

> *There is hope for improving the weather forecast*
> *Through a striking invention that has been made at last,*
> *Which is taking the place of the aeroplane flights*
> *To determine the state of the air at great heights.*

He describes the sending of data, saying, "Every fourth of a minute the signals come down." He laments the small number of balloons, which he attributes to their cost. He, like Lewis Fry Richardson and others, wonders when forecasters will work with the accuracy of astronomers. And though he writes of balloons, one stanza seems to foretell the capabilities of weather satellites armed with television cameras:

> *In the coming perpetual visiontone show*
> *We shall see the full action of storms as they go.*

*We shall watch them develop on far away seas,*
*And we'll plot out their courses with much greater ease.*

The complete collection of Mindling's *Weather Man Poems* is available online at http://www.history.noaa.gov/art/weatherpoems1.html.

The second US satellite to be put into orbit, Vanguard 1, was even smaller than Vanguard 2. It was a mere six inches in diameter and weighed three pounds (on earth). Launched on March 17, 1958, it drew the attention of Nikita Khrushchev, who reportedly referred to it as "the grapefruit satellite." The comparison was not meant as a compliment. Despite Khrushchev's sneering, his own satellite program was hardly above reproach. His second satellite, Sputnik 2, carried a dog into orbit in 1957. The dog was a stray picked up off the streets of Moscow and named Laika. At the time of the flight, the Russian authorities claimed that Laika lived six days in orbit, after which she was euthanized, but in 2002 it was revealed that the poor dog had died after only a few hours aloft, succumbing to the heat. A monument in Russia, erected in 2008, memorializes the little street dog who died in space.

Space junk, or orbital debris, is abundant. According to NASA, more than nineteen thousand pieces of space junk are tracked, and even more junk orbits untracked and unnoticed. Tracked space junk is at least the size of a marble. The typical travel speed is 17,500 miles per hour. Even a marble at that speed can ruin a space traveler's day. Twenty thousand pieces of space junk are larger than a softball. Collisions are rare but do occur. In 1996, space junk originating from a French rocket collided with a French satellite. In 2009, a dead Russian satellite collided with a live US Iridium commercial satellite, converting one piece of space junk and one live satellite into at least two thousand new pieces of space junk. In 2007, the Chinese tested an antisatellite weapon, creating another three thousand pieces of space junk. Some fireballs, like the one I saw from *Rocinante* crossing the sky toward Cuba, are bits of space junk reentering the earth's atmosphere. Most of the time, this reentering junk vaporizes before it reaches the ground.

Many people consider TIROS to be the first true weather satellite, but Vanguard 2 was considered a weather satellite when it went aloft in 1959, based on contemporary news accounts, including the one provided by Universal International News with the comforting narrative voice of the once well-known Ed Herlihy (http://www.youtube.com/watch?v=mgyhlQiKhhE). Many newspapers printed accounts of the successful TIROS launch on April 1, 1960. The words of Harry Wexler may seem odd to us today, when images from spacecraft have been commonplace for more than half a century, but in 1960 the success of TIROS must have seemed miraculous. For perspective, recall that a cell phone holds more computing power than was used to place TIROS in orbit. Also remember that in 1960, America's space program was best known for failures; exploding rockets were hardly newsworthy. The words of Wexler and the use of the word "moonlet" come from a UPI article published in many newspapers, including the *Honolulu*

*Star-Bulletin*. The article is available online at http://docs.lib.noaa.gov/rescue/ TIROS_newspaper_clippings_docs/19600401orbit_hsb.pdf.

## Chapter 5: The Numbers

The weather buoy off the coast of Florida was probably Station 42023-C13, owned and maintained, according to the National Data Buoy Center, by the University of South Florida.

The director of the Meteorological Office who commented on the value of weather maps was Robert Scott. "It is simply impossible that reports taken once a day can give an account of all the changes which supervene during the twenty-four hours," he wrote in 1875. "It is equally clear that to see at breakfast time a chart for the preceding morning is far less useful to us than a chart for the previous evening, not to say, even for midnight, would be." Scott attributed the shortcomings to inadequate funding. In the United States, funding also presented ongoing challenges, as it does today. Despite the apparent popularity of the weather map, personnel changes at the Weather Bureau and a paper shortage during World War I ended the regular appearance of weather maps in many newspapers in the United States. Citizens had to rely on brief written summaries or, by the 1920s, descriptions offered by radio stations. That ended in 1935 when the Associated Press launched its Wirephoto network, which delivered facsimiles by wire. Wirephoto allowed the AP to send both photographs and maps to subscribing newspapers around the country. Mark Monmonier's *Air Apparent: How Meteorologists Learned to Map, Predict, and Dramatize Weather* (1999, University of Chicago Press, Chicago) provides, among other things, a wonderfully detailed account of the history of the weather map in newspapers.

Bjerknes's description of his entry into meteorology as an interloper comes from a 1938 speech honoring the twenty-fifth anniversary of the Geophysical Institute of the University of Leipzig. An English version of the speech, translated by Lisa Shields, is available at Met Éireann, the Irish Meteorological Service, Dublin.

Richard Bach's *Jonathan Livingston Seagull* (1970, Macmillan, New York) was not only a bestseller but also something of a cult book that brought meaning to people feeling somewhat lost during an especially difficult period of American history. A friend of mine named his son Jonathan, after the gull, and he named his company (a diving company that I worked for as a teenager) Skybird Unlimited. While Bach saw the beauty in gulls, when I watch them today, especially in the tundra of northern Alaska, I often see them preying on the eggs and—even more heartbreaking—the downy young chicks of other birds.

The full text of Bjerknes's "Weather Predictions and the Prospect for Their Improvement," in English, can be found at http://folk.uib.no/ngbnk/Bjerknes_150/ bjerknes-1904-aftenposten-english.pdf.

The letters from Bjerknes to his friend describing his progress are quoted in a June 1981 article by Ralph Jewell called "The Bergen School of Meteorology: The Cradle of Modern Weather-Forecasting" (*Bulletin of the American Meteorological Society*, vol. 62, no. 6, pp. 824–30), available online at http://climate.envsci.rutgers.edu/pdf/JewellBAMS.pdf. The letters were to the Swedish scientist Svante Arrhenius, one of the early figures in the science of human-induced climate change. Arrhenius's work attempting to explain the ice ages led him eventually to write, "If the quantity of carbonic acid [$CO_2$] increases in geometric progression, the augmentation of the temperature will increase nearly in arithmetic progression." One letter from Bjerknes to Arrhenius said that funds had been allocated "for establishing a weather forecasting service for the whole of Western Norway."

For more on the history of broadcast meteorology, see Robert Henson's 2010 book *Weather on the Air* (American Meteorological Society, Boston). For short summaries about celebrities who were once forecasters, see http://www.mnn.com/lifestyle/arts-culture/photos/famous-people-who-used-to-be-weather-forecasters/forecasting-hot. The *TV Guide* article about the seriousness of broadcast forecasting appeared in the magazine's July 23, 1955, issue under the title "Weather Is No Laughing Matter." Francis Davis, who wrote the article, began his career in meteorology with the Army Air Corps and participated in the forecasts that supported the D-day invasion, an experience that must have influenced his view of forecasts as serious business. He was the first television broadcaster to receive the American Meteorological Society's Seal of Approval.

## Chapter 6: The Model

My 2013 book *Heat* (Little, Brown, New York) includes a chapter on the peat industry, focused on its historical importance. Prior to writing *Heat,* I had heard of peat being burned as a fuel, but I had never seen a brick of peat dried and ready for the fireplace. While writing the book, I acquired a block of peat, some of which I burned in my father-in-law's fireplace in Holland. He lives in the heartland of the Dutch peat-mining industry. When he was young, and especially in the midst of the coal shortages during and after World War II, his family regularly burned peat. When Lewis Fry Richardson was a young man, the peat industry was an important employer, and it is no more surprising to find that he once worked on peat management problems than it would be to see a young scientist today working for an oil company or a wind power company.

Lewis Fry Richardson, while not a name known in every household, is something of a cult figure in the world of meteorology. Lots of information and misinformation about his life and work are available from various sources. The description of Richardson's work as an ambulance driver near the front lines of World War I, along with other information about his life, comes from Oliver M. Ashford's excellent but little-known 1985 biography, *Prophet—or Professor?: The Life and Work of Lewis Fry Richardson* (Adam Hilger, Bristol and Boston). More readily available and also excellent but less detailed biographical information can be found in J. C. R. Hunt's 1998 article "Lewis Fry Richardson and His Contributions to Mathematics, Meteorology, and Models of Conflict," published in the *Annual Review of Fluid Mechanics* (vol. 30, pp. xiii–xxxvi) and available online at http://www.cpom.org/people/jcrh/AnnRevFluMech(30)LFR.pdf. The photograph of Richardson accompanying this short biography captures a sense of his intelligence, thoughtfulness, and reflective nature. Of the many people I have encountered while researching my books and articles, Richardson is near the top of the list of those I would love to meet, were he still alive.

Richardson's question regarding whether or not wind possesses velocity—that is, whether wind speed is a meaningful concept—comes from his paper "Atmospheric Diffusion Shown on a Distance-Neighbour Graph," published in the *Proceedings of the Royal Society of London. Series A, Containing Papers of a Mathematical and Physical Character* (1926, vol. 110, no. 756 [April 1], pp. 709–37). The third line of the paper is an equation with twenty-six terms.

Richardson was not the only Quaker to leave an important mark on meteorology. Another was Luke Howard, who, among other things, developed a nomenclature of clouds. His nomenclature is familiar to most people today even though it competed with the earlier one proposed by Jean-Baptiste Lamarck. Lamarck is best known for his ideas about evolution and his mistaken thoughts about the inheritance of acquired characteristics (for example, a horse stretching its neck to reach leaves on a tree passes along a stretched neck to its offspring, eventually resulting in giraffes). When not thinking about evolution (or other things), Lamarck identified five kinds (later twelve kinds) of clouds that he named in French. Howard used Latin names, such as *cirrus, cumulus, stratus,* and *nimbus.* Howard's work also attracted the attention of the poet Goethe, who wrote: "But Howard gives us with his clear mind, / The gain of lessons new to all mankind." He also inspired Percy Shelley's poem "The Cloud." Percy Shelley, a popular and influential poet, was married to Mary Shelley, the author of *Frankenstein.* Lamarck's classification scheme never had a chance against Howard's nomenclature. From the preface to the third edition of Howard's *Essay on the Modifications of Clouds* (1865, John Churchill and Sons, London): "From the time when this nomenclature was first suggested (about 1803), it has been universally adopted by scientific men, and, indeed, by all writers."

Barometric pressures as low as 870 millibars (25.7 inches of mercury) have been measured during hurricanes, and pressures even lower than that occur during tornadoes. Pressures as high as 1,086 millibars (32.0 inches of mercury) have been measured in strong high-pressure systems. However, the pressures predicted by Richardson's forecast, and the rate at which they changed, were outside of any normal expectations.

The book containing the copy of the ship's log entry describing sand three hundred miles off the coast of Africa is *The Sea Chart: The Illustrated History of*

*Nautical Maps and Navigational Charts,* by John Blake (2004, Naval Institute Press, Annapolis, MD). This is a beautiful book, available in paperback and full of color copies of charts and log entries. From time to time, and on those rare occasions when my charts mislead me, I look at this book to remind myself that there was a time when mariners sailed with charts that were both incomplete and wildly inaccurate. In comparison with mariners of the past, when ships were wood and men were steel, today's mariners appear overequipped, nervous, inept, and prone to unnecessary anxiety. Yet the death toll on the sailing ships of the eighteenth century was appalling, and with that in mind, I plan to acquire the most accurate charts available and to hold on to my anxiety for as long as I sail.

I came across the description of a soil scientist saying that farming practices in the Dust Bowl were "suicidal" in Timothy Egan's wonderful book *The Worst Hard Time: The Untold Story of Those Who Survived the Great American Dust Bowl* (2006, Houghton Mifflin, New York). Egan offers a gripping account of the history of the Dust Bowl, relying largely on firsthand accounts from survivors.

David Brunt's book is the second edition of *Physical and Dynamical Meteorology,* published in 1939 (Cambridge University Press, Cambridge). The first edition was published in 1934, and the book came out in paperback in 2011. Brunt focuses on theory and emphasizes mathematics, but he does not ignore fronts, sometimes in language that seems to echo Richardson. "For the moment we are only concerned to show that the system of prevailing winds brings into juxtaposition masses of air of widely different temperatures," he wrote, "and that these masses are separated by lines which appear as discontinuities on a chart of this scale."

The quotations from Sven Anders Hedin are from his 1944 book *History of the Expedition in Asia, 1927–1935,* vol. 3: *1933–1935,* published in Sweden but now available electronically. The book includes fascinating illustrations. Among many other things, Hedin describes dust storms: "The impression which such a tempest forces home upon you is, what an enormous deflation or power of transportation the wind must possess in a region such as this! All the fine dust which is produced through weathering and other agencies in the mountains of Central Asia, greatly disintegrated as they already are, all the dust which is produced by the friction of the particles of sand, grinding one against the other, and all that which originates from the mud deposited in the lakes which subsequently dry up—it is all borne westwards by these storms. Assuming that, during the stormy season, there is one black tempest a week, then each such storm carries away all the products of disintegration which have been accumulating throughout the preceding week and transports it a long way towards the west. After the cessation of one of these tempests it is quite noticeable how clean swept the surface appears; but it does not wear that appearance long, for the dust which is hovering in the air soon begins to settle again. At the end of a few days' calm, after a storm of this character, a layer of dust is deposited upon the sand, causing one's footsteps to stand out in lighter colouring."

David Cocks submitted his doctoral dissertation, "Mathematical Modelling of Dune Formation," to Lincoln College, University of Oxford, in 2005. I am entirely unqualified to judge the quality of the work, but nevertheless the dissertation leaves me deeply impressed. It is available online at http://eprints.maths.ox.ac.uk/764/1/cocks.pdf. Despite his mathematical focus, he includes a wonderful photograph of a sand dune in the Namibian desert and another of sand dunes on Mars, as well as numerous drawings illustrating aspects of dune formation and geometry.

Peter Lynch's fascinating reexamination of Lewis Fry Richardson's forecast is presented in his 2006 book *The Emergence of Numerical Weather Prediction: Richardson's Dream* (Cambridge University Press, Cambridge). Along with background information, Lynch runs Richardson's data through a modern computer, using the same mathematics employed by Richardson himself.

Richardson's famous error can also be explained by the failure of the continuity equation to account for short-term changes. In 1904, before Richardson's work, Max Margules assessed the usefulness of the continuity equation in predicting changes in pressure. He showed that the continuity equation would work only if wind data were known with a precision far beyond that which could be obtained in the practical world. In other words, he showed that the continuity equation would lead to failed forecasts. Like many before him, and some after him, Margules believed that forecasting was immoral and bad for the character.

Allyson Madsen and Jens P. Jeager owned the boat that was damaged in the Tortugas—a Whitby 42 ketch named *Indigo*. Their story is available at http://www.landfallnavigation.com/harrowing.html.

## Chapter 7: The Computation

The Dutch author is F. H. Klein, writing in *The Foretelling of the Weather in Connexion with Meteorological Observations,* originally published in English by Benjamin Pardon (1863, London) but available electronically today. The book's translator is listed as "A. Adriani, M.D., M.M., PH.D., F.R.N.I.E., F.R.S.S.A., Analytical and Professional Chemist; late Lecturer on Chemistry at the Medical College in connexion with Durham, University, Newcastle-on-Tyne, &c." We can all be grateful for the abbreviation "&c." The book also has a long subtitle, *Together with a Description of the Telegraphic Warning System Introduced in the Netherlands, June, 1860, as Proposed by the Director of the Royal Netherlands Meteorological Institute, Professor Dr. Buys-Ballot.* In the interest of holding the attention of readers and in the interest of my own sanity, while writing *And Soon I Heard a Roaring Wind* I was forced to ignore many stories and many great achievements. Although I think this is necessary for a book like this one, I wish I had been able to find a way to present the work of Christophorus Buys Ballot. Among other things, he gave us Buys Ballot's law: In the northern hemisphere, with your back to the wind, the low-pressure center stands to the left, an expression of the rightward motion of winds north of the equator.

Biologists have studied ballooning spiders in various ways. In some cases, nets sent aloft attached to kites or other aerial platforms are used to sample spiders from the sky. Other work is done in the laboratory. In a 1992 study by Robert B. Suter of Vassar College ("Ballooning: Data from Spiders in Freefall Indicate the Importance of Posture," *Journal of Arachnology*, vol. 20, pp. 107–13), unidentified species of small spiders were encouraged to hang from a strand of silk below a wire frame. A current sent through the frame severed the strand, allowing the spider to drop. A strobe fired one hundred times per second captured the spider's fall. "A spider falling through still air accelerates until the drag produced by air flowing past its body and trailing silk just equals the pull of gravity," wrote Suter. "At that point, acceleration is zero and the spider is at its terminal velocity." Surprisingly little is known about ballooning spiders. "Unfortunately," Suter wrote near the end of his paper, "almost nothing is known about the amount of silk actually used by ballooning spiders. That absence of data means that the importance of posture in the travels of ballooning spiders (even in their ability to control their elevation) cannot yet be assessed."

Swan wrote of the "aeolian zone" in 1961 ("The Ecology of the High Himalayas," *Scientific American*, vol. 205, no. 4, pp. 68–78) and again in 1963 ("Aeolian Zone," *Science*, vol. 140, no. 3562, pp. 77–78). By 1992, he had promoted his "zone" to a biome, as clearly articulated in his article "The Aeolian Biome: Ecosystems of the Earth's Extremes" (*BioScience*, vol. 42, no. 4, pp. 262–70). Swan presents a more extensive account of his work in the mountains in his 2000 book *Tales of the Himalaya: Adventures of a Naturalist* (Mountain N' Air Books, La Crescenta, CA), with a foreword by Sir Edmund Hillary. Swan taught at San Francisco State University for three decades. He also produced hundreds of television shows about science. According to one obituary, he determined that tracks in the snow attributed to Yeti were in fact the tracks of a mountain fox that hopped through the snow. The same obituary quoted him as saying that the Bayshore Freeway was more terrifying than the world's tallest mountains. Sadly, Lawrence Swan died the year before his book was released.

Dozens of sources describe Hurricane Gilbert and its impact on Cancún. One especially interesting source is an article by B. E. Aguirre called "Evacuation in Cancún During Hurricane Gilbert" (1991, *International Journal of Mass Emergencies and Disasters*, vol. 9, no. 1 [March], pp. 31–45). The words of the unnamed army officer are from the Associated Press article "Monstrous Storm Batters Yucatan," published on September 15, 1988.

The quotations from Landa's work are from William Gates's 1937 translation *Yucatan: Before and After the Conquest*. The translation was originally published by the Maya Society in Baltimore, which ran a total of eighty copies. Many of the illustrations were hand-tinted. My own copy is, alas, an electronic version of the later Dover edition, first published in 1978, with black-and-white illustrations. The book is worth reading for anyone traveling in or interested in the region. In the text, I refer to the friar as a murderer and book burner. He was also a torturer. All of this, of course, was in the name of religion and in the context of his time. "We found a large number of books in these characters," he wrote, referring to books written in the Mayan language, "and, as they contained nothing in which were not to be seen as superstition and lies of the devil, we burned them all, which they (the Maya) regretted to an amazing degree, and which caused them much affliction." Accused of managing an illegal Inquisition, Landa returned to Spain for trial. He was eventually absolved and appointed the second Bishop of the Yucatán. Although the extent of his cruelty may have been exaggerated over time, making Landa himself a victim of the so-called Black Legend, he was certainly not a kindhearted and merciful man by today's standards. As a footnote, William Gates's dedication illustrates the ephemeral nature of political change. He dedicated his translation of the book to Lázaro Cárdenas, the president of Mexico in 1937, crediting him with a solution to the "centuries-old problem of peasant farm lands" and with "rebuilding a nation of self-respecting and self-keeping citizens, and at the same time restoring the Indian race to a contributing position in the life of America."

Richardson was not alone among Quakers in standing by his principles and his religion despite the high costs. I recently came across other accounts of Quaker steadfastness in Peter Nichols's wonderful book *Oil and Ice* (2009, Penguin Books, New York). Nichols quotes the General Court of Massachusetts Bay Colony in 1657: "If any Quaker or Quakers shall presume, after they have once suffered what the law requireth, to come into this jurisdiction, every such male Quaker shall for the first offense have one of his ears cut off, and be kept at work in the house of correction till he can be sent away at his own charge, and for the second offense shall have his other ear cut off, and kept in the house of correction, as aforesaid." Quaker women guilty of the same offense were to be "severely whipped." As for third offenders, both men and women, "they shall have their tongues bored through with a hot iron." Nichols quotes another Quaker, Humphrey Morgan, in the relatively liberal (when it came to Quakers) colony of Plymouth, who had a few choice words for the colony's governor: "Thy clamorous tongue I regard no more than the dust under my feet and thou art like a scolding woman." In exchange for these words, he was fined and whipped.

In writing about big whirls and little whirls, Richardson was playing with the words of Augustus De Morgan in his 1872 book *A Budget of Paradoxes*. De Morgan wrote, "Great fleas have little fleas upon their backs to bite 'em, / And little fleas have lesser fleas, and so *ad infinitum*. / And the great fleas themselves, in turn, have greater fleas to go on; / While these again have greater still, and greater still, and so on." Morgan was himself playing with words that originated with Jonathan Swift's *On Poetry: A Rhapsody*, written in 1733: "So, naturalists observe, a flea, / Has smaller fleas that on him prey; / And these have smaller still to bite 'em, / And so proceed *ad infinitum*." Morgan was a mathematician and logician. Among other things, he coined the phrase "mathematical induction." Also from his book *A Budget of Paradoxes*, and likely of great interest to Richardson: "During the last two centuries and a half, physical knowledge has been gradually made to rest upon a basis which it had not before. It has become mathematical. The question now is, not whether this or that hypothesis is better or worse to the pure thought, but whether it accords with observed phenomena in those consequences which can be shown necessarily to follow from it, if it be true. Even in those sciences which are not yet under the dominion of mathematics, and perhaps never will be, a working copy of the mathematical process has been made." Meteorology found its way into Morgan's book: "Mr. Forster, an Englishman settled at Bruges, was an observer in many subjects, but especially in meteorology. He communicated to the Astronomical Society, in 1848, the information that, in the registers kept by his grandfather, his father, and himself, beginning in 1767, new moon on Saturday was followed, nineteen times out of twenty, by twenty days of rain and wind." Morgan used this anecdote as an example of coincidences that might appear to have a mathematical basis.

Richardson's first 1935 article in *Nature* was "Mathematical Psychology of War" (vol. 135 [May 18], pp. 830–31). From his abstract: "As *NATURE* has encouraged scientific workers to think about public affairs, I beg space to remark that equations, describing the onset of the War, and published under the above title in 1919, have again a topical interest, in connection with the present rearmament." Even the abstract includes mathematics. In a follow-up letter to *Nature*, with the same title but published seven months later (vol. 136, p. 1025), he wrote: "The international situation is thus represented by a point $(x, y)$ in a plane. Let us think of this point as a particle moving in accordance with the equations. If the particle be tending towards plus infinity of both $x$ and $y$, then war looms ahead. But if the particle be going in an opposite direction the prospect is peaceful." Both papers grew from the 1919 book mentioned in the first 1935 *Nature* article, which, like the book, was titled "Mathematical Psychology of War." The book was dedicated to his "comrades of the motor ambulance convoy." In the introduction, he explains in part his love of mathematics. "To have to translate one's verbal statements," he wrote, "into mathematical formulae compels one carefully to scrutinize the ideas therein expressed. Next the possession of formulae makes it much easier to deduce the consequences. In this way absurd implications, which might have passed unnoticed in a verbal statement, are brought clearly into view and stimulate one to amend the formulae." Not surprisingly, readers, including Richardson's friends, found his approach tough going, but Bertrand Russell was among his supporters.

Eisenhower's adviser was James Martin Stagg. His experience and words were later detailed in his book *Forecast for Overlord* (1972, W. W. Norton, New York). As is often the case, more than one interpretation of the discussions described by Stagg is possible, but only one interpretation is presented here for simplicity. Operation Overlord was the code name of the invasion of Normandy. Operation Neptune, as the D-day landings were called, was part of Operation Overlord.

The two key forecasters for the D-day invasion were the American Irving Krick and the Norwegian Sverre Petterssen. Krick claimed credit for the forecast that allowed the invasion to go forward despite the fact that his forecast would have sent the troops across the English Channel in a gale. James Martin Stagg was tasked with filtering the information that came from Krick and Petterssen and resolving the differences. After the war, Krick continued in weather forecasting but also went on to establish a cloud-seeding, rainmaking business with more than one hundred employees. In his youth, before the war, he was a musician, and it was later said that he came to weather forecasting only because it paid more than piano playing.

The words of Jule Charney are from his 1949 paper "On a Physical Basis for Numerical Prediction of Large-Scale Motions in the Atmosphere" (*Journal of Meteorology*, vol. 6, no. 6, pp. 372–85). Charney left many marks on meteorology, including a short report titled "Carbon Dioxide and Climate: A Scientific Assessment" (1979, National Academy of Sciences, Washington, DC). He was the lead author of the Ad Hoc Study Group on Carbon Dioxide and Climate, which wrote the report. Although the fact that atmospheric carbon dioxide traps heat was known in the 1800s, and although the possibility that emissions from fossil fuels could increase carbon dioxide levels sufficiently to affect climate had been discussed for nearly a century (a reality first posited by Svante Arrhenius, who was a frequent correspondent and supporter of Vilhelm Bjerknes), the 1979 report offered a numerical estimate of the relationship between carbon dioxide levels and temperature: "When it is assumed that the $CO_2$ content of the atmosphere is doubled and statistical thermal equilibrium is achieved, the more realistic of the modeling efforts predict a global surface warming of between 2°C and 3.5°C, with greater increases at high latitudes." The word "wind" appears three times in the report. It seems that wind is rarely discussed in the climate change literature even though wind is driven in large part by differences in temperature between the higher latitudes and the tropics. The report is available online at http://web.atmos.ucla.edu/~brianpm/download/charney_report .pdf. I recommend reading it in its entirety.

Mark Vanhoenacker's 2015 book *Skyfaring: A Journey with a Pilot* (Alfred A. Knopf, New York) offers, among other things, a commercial pilot's view of moving air, including jet streams. To those of us who fly as passengers, the beauty of flight is often lost in tightly packed seats and the generally uncomfortable reality of air travel, but Vanhoenacker's personal insights help readers see beyond the immediate discomfort and allow them to share his love of flying.

Rossby waves can be understood (and misunderstood) on multiple levels. Rossby explained them and understood them mathematically, a feat that is beyond the grasp of most mortals. In the text, I explain them through analogies, and analogies are never perfect. An expository explanation could have started by saying that Rossby waves are really just large-scale, high-altitude meanders in winds that emerge because the influence of the Coriolis effect changes with latitude, leading to shearing between air masses, which is itself affected by inertia. Clearly, the explanation by analogy will satisfy more readers than the explanation by exposition.

Jule Charney's paper "On the Scale of Atmospheric Motions" was published in *Geofysiske Publikasjoner* (1948, vol. 17, pp. 1–17). The Norwegian influence is present even in Charney's work, but thankfully he wrote in English. His paper is available online at http://empslocal.ex.ac.uk/people/staff/gv219/classics.d/Charney48.pdf.

Jule Charney, with coauthors Ragnar Fjörtoft and John von Neumann, summarized the ENIAC forecasts in a 1950 article wonderfully titled "Numerical Integration of the Barotropic Vorticity Equation" (*Tellus,* vol. 2, pp. 237–54). "It may be of interest," they wrote, "to remark that the computation time for a 24-hour forecast was about 24 hours, that is, we were just able to keep pace with the weather. However, much of this time was consumed by manual and I.B.M. operations, namely by the reading, printing, reproducing, sorting, and interfiling of punch cards." The article includes undecipherable (for me) equations, as well as maps comparing forecasts and realities. About a third of the article is devoted to a discussion of the things that went wrong. "An attempt," they wrote, "will now be made to account for the errors in the forecasts." The full text is available online at http://mathsci.ucd.ie/~plynch/eniac/CFvN-1950.pdf.

## Chapter 8: Chaos

Darwin saw atolls at Fiji and in the Cocos Islands. Captain FitzRoy sampled the depths, trying to understand the extent of the living coral. Darwin's hypothesis about coral atoll formation suggested that the reef should be thick. As the island sank and the coral grew, staying close to the surface, the coral reef would become thicker and thicker. If the reef was nothing more than a thin veneer, as was often thought, a drill bit penetrating into the coral would soon hit sand or, at least, something other than coral. If the reef was growing as the island subsided, the drill bit would turn up coral, and more coral, and more coral. Coral atoll drillers, working between 1896 and 1898 at the Funafuti Atoll, penetrated just beyond 1,114 feet. All of the cuttings coming up from the hole were coralline. In the 1950s, as part of the preparations around hydrogen bomb testing at Eniwetok, drillers penetrated past 4,000 feet, finally reaching underlying basalt.

Charles F. Brush's giant (for its time) wind turbine reportedly generated 12 kilowatts of power. Aboard *Rocinante,* in addition to the 120-horsepower diesel that powers the seldom-used propeller, we carry a 9-kilowatt generator. On *Rocinante,* we mainly run the generator to keep its engine's cylinders from rusting.

T. Lindsay Baker's 2007 book *American Windmills: An Album of Historic Photographs* (University of Oklahoma Press, Norman) includes dozens of black-and-white photographs showing windmills and their uses. Thumbing through this book (a highly recommended activity) will leave one with the correct impression that windmills were once a common sight on the American landscape.

YouTube offers videos of failing wind turbines. For example, for a turbine that failed catastrophically in high winds, see http://www.youtube.com/watch?v =u14tBwO5QVQ, or for a turbine that ignited, see https://www.youtube.com/watch?v=aegHUv2OkEE. Unrelated but entertaining, watch a BASE jumper leap off a turbine blade at http://www.youtube.com/watch?v=jQZ_PhvRE14, bearing in mind that BASE jumping remains one of the most dangerous sports (or stunts) in the world.

T. Boone Pickens's words about not turning green were reported in the April 13, 2008, issue of the *Guardian*. His response to the woman's question about turbine noise appears in the thorough and readable modern history book *The Great Texas Wind Rush,* by Kate Galbraith and Asher Price (2013, University of Texas Press, Austin). His words about losing money in the wind business come from an interview on MSNBC's *Morning Joe* on April 11, 2012. In 2013, Pickens fell off the Forbes 400 list of the richest Americans, putatively because of investments in wind that left him practically destitute, with a net worth of "around $950 million." At the time of this writing, he is again a billionaire.

Thoreau's words about wind power come from his "Paradise (to Be) Regained," published in the November 1843 issue of the *United States Magazine and Democratic Review* and later collected in *A Yankee in Canada, with Anti-slavery and Reform Papers* (1866, Ticknor and Fields, Boston). Abraham Lincoln's words on the power of wind come from his "Discoveries and Inventions" lecture in 1860. I came across an account of this lecture in Jeremy Shere's 2013 book *Renewable: The World-Changing Power of Alternative Energy* (St. Martin's Press, New York).

The German inventor of the original radar—really a pre-radar that demonstrated the principles of radar without measuring distances—was Christian Hülsmeyer. His telemobiloscope, as it was called, could not estimate distances. It said only, in essence, "There's something out there." Hülsmeyer apparently recognized the value of his device in shipping. He patented it under the title (in English) "Hertzian-Wave Projecting and Receiving Apparatus Adapted to Indicate or Give Warning of the Presence of a Metallic Body, Such as a Ship or a Train, in the Line of Projection of Such Waves." He attracted an investor—a leather merchant—and formed a company. Later, with different investors, his device apparently failed. He moved on to other projects. The translated title of the newspaper article suggesting that the telemobiloscope might have military applications is "Ship Collision Avoidance Instrument," published on June 11, 1904, in the Dutch newspaper *De Telegraaf.* More details on Hülsmeyer and his invention can be found in many sources, including a 2002 speech given in his honor, which is available online at http://www.design-technology.info/resource documents/Huelsmeyer_EUSAR2002_english.pdf.

Joshua Slocum, like many sailors, was a reader. Among the many books he carried aboard was *Don Quixote.* The last reported sighting of Slocum was in 1909, when he set off once again alone. His wife believed he was dead by 1910. He was not declared legally dead until 1924. As my faith in the authorities is somewhat limited, I still hope to see him one day, aboard his famous boat *Spray*, sailing before a fair wind into a setting sun. I suspect I am not the only sailor who harbors this hope.

The words of F. O. Willhofft, who had previously worked as a professor of mechanical engineering at Columbia University, are from the *New York Times*, May 3, 1925. He was also quoted as saying, "The outstanding fact is that rotating cylinders produce about ten times the propulsive force as canvas sails of the same area and that the actual results obtained in the trial trips of the *Buckau* confirmed the laboratory results with remarkable exactness."

For an interesting video showing the *E-Ship 1* under way, with its Flettner rotors spinning, see https://www.youtube.com/watch?v=2pQga7jxAyc.

The Beluga Shipping executive who felt sailing in his bones was CEO Niels Stolberg. His words were reported in a Sailing-World.com article by Roger Boyes published on December 15, 2007. Beluga Shipping's insolvency appears to have resulted from the global recession. Unrelated to the *Beluga SkySails*, Beluga Shipping was accused of illegally transporting arms to Myanmar and South Sudan, but the vessels in question were under lease at the time. Stolberg was cleared of all charges. To my knowledge, neither of the ships allegedly involved in illegal activities used the giant kites produced by the German company SkySails. That company remains in business with, according to its website, about fifty employees.

Historically, most ambergris comes from hunted whales. With that in mind, many nations have passed regulations outlawing the ambergris trade, even ambergris found on a beach or afloat at sea, well removed from the whale that made it.

The description of skill scores, as well as much of the background I have used throughout this book, is from Mark Monmonier's aforementioned *Air Apparent: How Meteorologists Learned to Map, Predict, and Dramatize Weather*. Monmonier is a geographer, fascinated by maps and mapping, and is also the author of *How to Lie with Maps* (2nd ed., 1996, University of Chicago Press, Chicago) and a string of other titles that bring the nuances of cartography to the shelves and minds of ordinary readers.

Edward Lorenz's famous "Predictability: Does the Flap of a Butterfly's Wings in Brazil Set Off a Tornado in Texas?" is available online at http://eaps4.mit.edu/research/Lorenz/Butterfly_1972.pdf and elsewhere. The paper, which was delivered to the American Association for the Advancement of Science on December 29, 1972, is both worth reading and very readable. The title, which came as a suggestion from a conference organizer familiar with Lorenz's work, is brilliantly appealing, but it may have led some to believe that one small event can lead predictably to another—that is, that the flapping of a particular butterfly's wings at a particular place may predictably lead to a particular tornado at some future date. That is, of course, precisely not Lorenz's meaning. Movies and novels flashing back to a single trivial event that predictably shapes an entire future misrepresent his work. Fate, as the gods would have it, is not that simple.

Anyone with a computer and a bit of spreadsheet experience can see for himself or herself how very small differences in initial conditions lead to large and unpredictable differences in the future using simple equations. The simplest and most accessible equation is probably the logistic equation, sometimes used to model population growth. With low values of the equation's rate constant, results behave themselves, becoming stable after some period (or, equivalently, some number of reproductions), representing a stable population. As the rate constant increases, the equation gives oscillating results, so that the population fluctuates regularly between two levels. A further increase in the rate constant leads to chaotic results, without any stability. Also, minor differences in initial conditions lead to minor differences in future results as long as the rate constant is low, but with a rate constant high enough to lead to chaotic results, minor differences in initial conditions lead to major differences in future results, just as one would expect with chaotic systems.

### Chapter 9: The Ensemble

Edward Lorenz's 1964 paper discussing error rates (and many other things) is "Large-Scale Motions of the Atmosphere: Circulation," published in *Advances in Earth Science: Contributions to the International Conference on the Earth Sciences,* edited by P. M. Hurley (1964, MIT Press, Cambridge, MA, pp. 95–109), available online at http://eaps4.mit.edu/research/Lorenz/Large_Scale_Motions _of_the_Atmosphere_1966.pdf. In this paper, he alludes to chaos theory but does not dwell on it. At one point, in what to me is a striking sentence, he talks about determinism without mentioning chaos. "We shall not be so much concerned with the philosophical question of determinism," he wrote, "or whether the atmosphere has decided what to do, as with whether it has signaled its intentions to us."

After talking to the other sailors in Placencia, I checked on the well-known aphorism about the difficulty of predicting the future. According to Quote Investigator (http://quoteinvestigator.com/2013/10/20/no-predict/), the words can be traced back to at least the 1930s, when they were used in the Danish parliament. By 1956, the words, more or less, appeared in the *Journal of the Royal Statistical Society, Series A.* In 1971, and several times thereafter, they were attributed to Niels Bohr. In 1991, a publicist credited Yogi Berra with them. Quote Investigator does not cite anything written by Bohr or Berra himself that actually contains these words. When it comes to attributions of catchy phrases, it appears that it is just as difficult to make predictions about the past as it is to make predictions about the future.

NOTES

The newspaper report on Operation Popeye was distributed by the New York Times News Service in May 1974. Operation Popeye relied on cloud seeding. The report notes that officials did not agree on the effectiveness of the program, with some claiming a thirty percent increase in rainfall and others claiming as little as a couple of additional inches in areas already receiving, without military intervention, twenty-one inches of rainfall. The beauty of rainmaking in the tropics is that almost any claim of success can be defended. Another widely discussed approach to weather control (that is, among conspiracy theorist crowds) points to the HAARP (High Frequency Active Auroral Research Program) facility in Gakona, Alaska. In reference to the facility's cost, *Wired* magazine commented in 2009 that "the Pentagon wanted to know when its overpriced conspiracy-magnet would produce that battle-ready technology they'd been promised." The politically connected activist Nick Begich and coauthor Jeane Manning wrote *Angels Don't Play This HAARP* (1995, Earthpulse Press, Eagle River, AK), which includes discussion of weather control. The 1996 US Air Force report is *Weather as a Force Multiplier: Owning the Weather in 2025*, available online at http://csat.au.af.mil/2025/volume3/vol3ch15.pdf. In researching weather weaponization schemes and discussions, it struck me repeatedly that much of what is available to the public is so ridiculous that it may be a form of unsophisticated counterintelligence planted to fool the enemy, whether the enemy is real or imagined. In any case, Cynthia Barnett's *Rain: A Natural and Cultural History* (2015, Crown, New York) offers a fascinating summary of the history of weather control for military and other purposes.

The pressure, or force, generated by wind can be approximated with simple formulas or from simple tables. Force calculations include a squared term for wind velocity. This means that a doubling of wind speed increases wind force four times, a tripling of wind speed increases wind force nine times, and so on. Even though the relationship between wind speed and wind force seems straightforward, calculating the actual force of the wind lies somewhere between difficult and impossible. Wind speed changes all the time, and the force from a gust is much stronger than the force of the average wind. Also, wind at height—near the top of the mast—is usually stronger than wind near the water or the earth's surface (where friction slows everything down). Nonlaminar flow—that is, turbulent flow, exactly the kind of wind movement that occurs near anything offering resistance—complicates the simple relationship between wind speed and wind force. For boats, add to all of this the complication of heeling—the boat leaning over as the wind strengthens—which changes the cross-sectional area of the surface on which the force can act. And, of course, the boat may move as wind speed increases, and any motion of the boat downwind has to be subtracted from the measured wind speed. I could go on, but it just gets worse and worse.

When Lorenz ran his initial models, the computer calculated a new answer for every six hours represented in the model. To run four six-hour increments—that

297

is, to run one day's worth of "weather" on a computer that was, compared with modern computers, little more than an abacus—required about one minute of computer time. Printing the results slowed things down even more. Lorenz printed results from every fifth model run, or every thirty hours.

*Desperate Voyage,* first published in 1949 but reprinted in 1991 (Sheridan House, New York), is John Caldwell's account of his Pacific crossing. After being discharged from the military, Caldwell found himself in Panama. He hoped to rejoin his wife in Australia, but in the troubled postwar world, transportation was scarce. Caldwell bought a small wooden yacht. Upon seeing the yacht, Caldwell's sailing partner backed down. Caldwell set off on his own, learning to sail as he went. With no weather forecasts, he encountered storms. His damaged boat leaked. He pumped the bilges frequently. He ran out of food. He persevered. Eventually, he grounded the damaged boat on an island. By the time he reached the island, he weighed about ninety pounds. Books like *Desperate Voyage* are fun to read, but they also offer perspective for modern yachtsmen, reminding all of us that the number one function of a boat, and in the end possibly its only important function, is to remain afloat. The expensive navigation gear, the teak trim, the flushing head, and all the little extras that drive up the cost of sailing are not the most important elements of a sailing vessel.

## Chapter 10: Afloat in the Candle's Light

The 2007 interview of Edward Lorenz, by Robert W. Reeves, appeared in the May 2014 issue of the *Bulletin of the American Meteorological Society* (vol. 95, pp. 681–87). The interview was, according to an American Meteorological Society abstract, "part of a larger effort to document the history of operational long-range prediction," which may explain why it was not published until well after Lorenz's death in 2008. The interview transcript is available online at http://journals.ametsoc.org/doi/pdf/10.1175/BAMS-D-13-00096.1.

A video—a cartoon, really—illustrates the Madden-Julian oscillation, showing how weather in the Indian Ocean moves across the planet. It is available online at http://www.ucar.edu/communications/video/dynamo.mov?_ga=1.158325882.1 532294756.1428078794.

Republican congressman Jim Bridenstine spoke to his colleagues about the relative costs of climate change research and forecasting on June 11, 2013. His words were reported by C-SPAN and others, and were checked by the *Tampa Bay Times*'s PolitiFact, http://www.politifact.com/truth-o-meter/statements/2013/jun/ 14/jim-bridenstine/rep-jim-bridenstine-says-us-spends-30-times-much-c/. PolitiFact reporters and researchers spend their days fact-checking the statements of politicians. According to PolitiFact, Bridenstine said that his comments were about forecasting research, not forecasting as a whole. Nevertheless, PolitiFact ranked his comparison as "mostly false" on their trademarked Truth-O-Meter.

In September 2009, *Scientific American* ran an article online called "Climate Change May Mean Slower Winds," a title that says it all. The article is available at http://www.scientificamerican.com/article/climate-change-may-mean-slower -winds/. Diandong Ren's "Effects of Global Warming on Wind Energy Availability," published in the *Journal of Renewable and Sustainable Energy* in 2010 (vol. 2), also predicts decreased average wind speeds with increased global temperatures. It is available at http://scitation.aip.org/content/aip/journal/jrse/2/5/10.1063/1.3486072.

The *Scientific American* article by Jeff Masters, "The Jet Stream Is Getting Weird," appeared in December 2014 (vol. 311, no. 6, pp. 68–75). Masters did a masterful job of explaining the effect of climate change on Rossby waves and the effect of Rossby waves on the tough winters that hit the Eastern Seaboard the year his article appeared.

University of Washington professor Cliff Mass describes some of the complaints of meteorologists in the blog post "The U.S. Has Fallen Behind in Numerical Weather Prediction: Part I," *Cliff Mass Weather Blog,* http://cliffmass .blogspot.com/2012/03/us-fallen-behind-in-numerical-weather.html, noting that they are not his alone but can also be found in committee reports. Mass, in his many fascinating blog posts, does not come across as a complainer, but rather as an energetic and tireless researcher who counts crowdsourcing of weather data among his many interests. His description of these complaints is best seen as constructive criticism. He and probably all of his colleagues sincerely want to see improved forecasting.

# INDEX

Adams, John Quincy, 74
*Admiral FitzRoy* (ship), 100
air. *See* atmosphere
Air Commerce Act (1926), 125
aircraft, 122–25
    data collection by, 104, 105–6,
        124–25, 152–53
    experimental rotor, 211
    forecasts for, 262–63
air mass analysis, 145, 146
air pollution, 169–70, 176
AIS (Automatic Identification
        System), 71, 72
Alberti, Leon Battista, 95, 97
Alcott, Louisa May, 16
*Alcyone* (ship), 209–10
ambergris, 212–13
ambulance drivers, World War I,
        141–42, 178
AMDAR (Aircraft Meteorological
        Data Relay), 105–6
American Association for the
        Advancement of Science, 75
American Meteorological Society, 131
America's Cup race, 32–33
anemometers, 29, 86, 93, 94–98,
        138, 256
animals
    forecasts by, 165–66
    marine, 93–94, 162–64
    plants and, 168–69
    *See also* birds; insects
*Ann McKim* (ship), 225

Arago, François, 47, 64
Archimedes, 66, 67, 77, 146
astrological meteorology, 60–61
atmosphere
    air movement in, 66–67, 69–70,
        73–75, 80, 81–83, 91–92, 138,
        265
    dynamic balance in, 158
    gravity waves in, 158–59
    layers of, 105, 137
    natural laws for, 78, 79, 128,
        240, 251
    predicting behavior of, 119–20
    water compared to, 19–20, 66
    weight of, 89–90, 92
atmospheric pressure, 20
    Ferrel on, 82–83
    hurricanes and, 30
    mapping changes in, 37, 112
    measurement of, 86, 87–93, 252
    Richardson on, 142–43, 157, 160
    wind and, 91–92
Automatic Identification System
        (AIS), 71, 72
aviation. *See* aircraft

Bach, Richard, 121–22
Ball, John, 41–42
balloons, 104–5, 127, 140, 157
*Barometer and Weather Guide*
        (FitzRoy), 92–93
barometers, 86, 87–94
    *See also* atmospheric pressure

*Beagle* (ship), 33–34, 48, 50, 108, 130, 167, 194

Beaufort, Sir Francis, 33–34
  telegraph and, 45–46
  wind scale of, 26, 27–29, 94, 256

*Beluga SkySails* (ship), 211–12, 214

Bengtsson, Lennart, 10

Bergen School, 127, 132, 133, 180
  methods of, 144–46, 153, 160, 184, 185

Berra, Yogi, 230

Berti, Gasparo, 88, 89

Bertrand, Abbé, 44

Bilas, Fran, 186

biomes, aeolian, 167–70

birds, 13, 121–22, 125, 162, 165

Bjerknes, Jacob, 132

Bjerknes, Vilhelm, 116–21, 133–34, 251, 263, 266
  Charney and, 190–91
  data compiled by, 140
  death of, 133
  forecasting by, 84, 109, 119–21, 125–29, 132, 135–36, 143, 176, 205, 227
  fronts and, 115, 181
  methods of, 144–46, 160, 185

Board of Trade, British, 47–48, 50, 92

Bohr, Niels, 230

Bonaparte, Louis-Napoléon, 45

Bonaparte, Napoléon, 45

Bonny, Anne, 208

Boswell, James, 106

Boyle, Robert, 90

Brahe, Tycho, 28

British Association for the Advancement of Science, 58

Brunt, David, 153

Brush, Charles F., 196–97, 201

*Buckau* (ship), 210–11

buoys, 102–3, 112, 156

Bush, George W., 199

Caldwell, John, 241–42

*Captive* (weather balloon), 104–5

Carnegie Institution, 127, 146

*Celestial Mechanics* (Laplace), 77–78, 226

cell phones, 252, 255

Celsius, Anders, 28

centrifugal force, 30–31, 81

Chandler, Raymond, 23

chaos theory
  forecasting and, 34, 220–23, 226–29, 235–40, 244, 257
  weather control and, 231, 232

Charney, Jule, 227, 266
  ENIAC and, 186–87, 189–91, 222, 230, 263

chlorine, 176–77

Churchill, Winston, 180

circulation theorem, 118–19, 128

climate change, 259–61

Cocks, David, 155

Columbus, Christopher, 23, 24–25, 26

computers
  air pollution models in, 176
  ENIAC, 134, 186–87, 189–92, 218, 222, 230, 255, 263
  forecasting by, 11–12, 53–54, 156, 185–92, 207, 217–23, 234–35, 242–44, 255–56
  human, 10–11, 185–86
  MANIAC, 230
  PHONIAC, 255

coral reefs, 194–95, 204

Córdoba, Francisco Hernández de, 174

Coriolis, Gaspard-Gustave, 70, 72, 77, 81, 158, 188

Cousteau, Jacques, 209

Crawford, William, 181

Crimean War, 43–45, 50, 241

*Cutty Sark* (ship), 225

cyclones, 24, 42
  *See also* hurricanes

cyclostrophic balance, 31

Darwin, Charles, 61
  *Beagle* voyages of, 34, 48, 49–50, 59, 66, 108, 166–67, 194–95
  theory of, 58, 59, 63, 163–64

da Vinci, Leonardo, 95
Defoe, Daniel, 3–8, 13, 19, 34, 266
Delahaye, Jacquotte, 208
depth sounding, 110–11
derecho, 23–24
*Desperate Voyage* (Caldwell), 241–42
determinism, 232, 240
Dickens, Charles, 53
Doppler radar, 206–7, 217
Doré, Gustave, 1
Doyle, Arthur Conan, 209
Duckworth, Joe, 152
Dust Bowl, 134, 148–53, 155

Einstein, Albert, 117
Eisenhower, Dwight D., 183, 184–85
*Elizabeth Watts* (ship), 213, 214
Ellison, Larry, 33
energy, wind, 13, 196–97, 199,
    200–204, 211
ENIAC (Electronic Numerical
    Integrator and Computer), 134,
    186–87, 189–92, 218, 222, 230,
    255, 263
*E-Ship 1* (ship), 211
Espy, James, 72–77, 83, 261
*Essay on the Winds and the Currents of
    the Ocean, An* (Ferrel), 78–79,
    81–83
ether, 117, 118, 119
eugenics, 62
Euler, Leonhard, 67, 77, 146
Evans, Lewis, 39
evolution, theory of, 58, 59, 63, 163–64

Farrell, Sadie "the Goat," 208
Ferrel, William, 77–79, 81–83, 84, 86,
    226, 266
Ferrel cells, 69–70
fish, flying, 93–94
FitzRoy, Robert, 48–51, 76, 251, 266
    animal forecasting and, 165–66
    barometer and, 92–93
    Bjerknes and, 126
    critics of, 60–64, 104, 255, 261
    Darwin and, 34, 48, 49–50, 59, 66,
        108, 130, 166, 194

forecasting by, 55–60, 66, 84, 133,
    207
    suicide of, 64, 78, 117
    telegraph and, 55–56, 86, 94
Fletcher, H. M., 200, 214
Flettner, Anton, 210–11, 214, 217, 266
forecasting. *See* weather forecasting
*Foretelling Weather* (Saxby), 60,
    116–17
Franklin, Benjamin, 38–40, 42, 73,
    74, 105
Franz Ferdinand, Archduke of
    Austria, 140
fronts, 34, 132, 181
    forecasting models for, 145–46
    mapping, 115, 129–30, 146, 153

Galileo, 87–88, 89, 92
Galton, Francis, 61–62
Galveston, Texas, 3, 15–16, 17–18
gas warfare, 177–78, 180, 181
geostrophic wind, 83
GFS (Global Forecast System), 53, 54
Glaisher, James, 62, 104–5
global warming, 259–61
Goad, John, 61
GOES-13 (Geostationary Operation
    Environmental Satellite 13), 252
gravity waves, 158–59
Great Plains, drought in, 148–53
grid
    Charney's, 190, 191
    Richardson's, 41, 53, 136–37, 140
gusts, 34, 246

Hadley, George, 68–70, 77, 80,
    81–82, 188
Hadley cells, 69
Halley, Edmond, 38, 67–68, 98, 250
heat waves, predicting, 258
Hedin, Sven Anders, 154
heliostat, hand, 62
Helmholtz, Hermann von, 117, 119
*Henri IV* (ship), 44, 45, 47, 50
Henry, Joseph, 46, 76, 94
Herschel, Sir John, 61, 63
Herschel, William, 61

Hooke, Robert, 28, 90, 91, 95
Horace, 165
horse latitudes, 80–81, 82
HRRR (High-Resolution Rapid
    Refresh) models, 257–58
hurricanes, 24–25, 29–31
    categories of, 28, 172
    extratropical, 42
    eye of, 30, 31
    flying into, 152
    in Galveston, 3, 15–16
    Gilbert, 172–73
    season for, 26
    Wilma, 175, 195
Huxley, Thomas, 58, 59
hydrodynamics, 66, 117, 118–19, 128
hydrofoils, 33
hydrostatics, 66

Iger, Bob, 130
initial conditions, 84, 98, 113, 120,
    251–52
insects, 163, 166–67, 168–69
isobars, 37, 112

Jennings, Betty Jean, 186
jet streams, 187–88, 260
*Jonathan Livingston Seagull* (Bach),
    121–22

Kelvin, Lord, 119
*Kenilworth* (ship), 44, 241
Kepler, Johannes, 60–61
Kern, Lawrence, 3
kites, 127, 212, 214
Kitty Hawk, 122–23, 124

lag deposits, 154, 155
Landa, Friar Diego de, 174–75, 193
landscape, reshaping, 153–55
Laplace, Pierre-Simon, 77–78, 81,
    226–27, 240, 243
lead straw, water-filled, 88–89
lee shore, 198–99, 207
Letterman, David, 130
Le Verrier, Urbain, 44–45, 47, 86, 94
Li, Tien-Yien, 226

Lincoln, Abraham, 203, 204
Loopers, 129
Lorenz, Edward, 218–22, 251, 266
    chaos theory and, 220–22, 226,
        228–29, 236–39, 244, 257
    death of, 244
    Laplace's demon and, 227
lunarism, 60–61, 117, 133
Lynch, Peter, 157–58

Madden-Julian oscillation, 258
Magnus effect, 210, 211
MANIAC (Mathematical Analyzer,
    Numerical Integrator, and
    Computer), 230
man-of-war (ship), 27–28, 164
man-of-war, Portuguese (animal), 27,
    163–64
maps, 38–40
    thematic, 38
    trade wind, 38, 67–68, 98, 250
    *See also* weather maps
Masters, Jeff, 260–61
mathematics, 134–40
    of air movement, 81–83
    application of, 8–9, 77, 78, 119–21,
        127, 218–19, 220
    differential equations in, 134–36,
        137, 182
    finite difference method in, 134–35,
        137, 254
    strange attractor in, 238–39, 240
    theory combined with, 78, 84
    *See also* weather forecasting
Maury, Matthew Fontaine, 78
Mayans, 173–74
Melville, Herman, 198, 212
mercury, 89–91
Meteorological (Met) Office (UK), 48,
    56, 99, 114
meteorologists, 37, 77, 129–32, 153,
    180, 259, 261
microwaves, 109
*Moby-Dick* (Melville), 198, 212
momentum, linear and angular, 70,
    81, 158
Morse, Samuel, 45, 46, 86

NAM (North American Mesoscale
    Forecast System), 54
National Weather Service, 206–7
    *See also* Weather Bureau, US
NCEP (National Centers for
    Environmental Prediction),
    53, 54
neuston, 162–63, 168
newspapers, 15–16
    Dust Bowl stories in, 150, 151
    forecasts and maps in, 56–57,
        63–64, 112–15
Newton, Isaac, 8, 48, 67, 73, 77, 78
NOAA (National Oceanic and
    Atmospheric Administration), 53,
    54, 101
NOMADs (Navy Oceanographic
    Meteorological Automatic
    Devices), 102–3
northers, 25–26, 36–38, 42

Obama, Barack, 259
observation, theory and, 76, 78, 109,
    127, 133
occlusion, 181
O'Hair, Ralph, 152
*On the Origin of Species*
    (Darwin), 58
optical scanners, 106
Osborne, Michael, 201, 202

Pain, Derick, Jr., 7
pans, 154, 155
parameterization, 254–55
Parker, Chris, 229, 266
PassageWeather (website), 53, 54,
    234–35, 266
peat bogs, 134, 137
Pennsylvania, map of, 39
*Philosophy of Storms, The* (Espy),
    73–74
PHONIAC (Portable Hand-Operated
    Numerical Integrator and
    Computer), 255
physics, laws of, 8, 119–21, 127,
    218–19, 220
Pickens, T. Boone, 202–3

pirates, 208
plants, 148–49, 168–69
Pliny the Elder, 165
Pocahontas, 25
polar cells, 70
*Polarfront* (ship), 101
positivism, 78
Post, Wiley, 187
PressureNet, 252
*Prince* (ship), 44, 47, 50,
    241, 245
*Progress* (ship), 44, 50, 241
Ptolemy, 60

quasi-geostrophic (QG) theory, 189

radar, 204–7, 217
Radner, Gilda, 130
Read, Mary, 208
Redfield, William C., 73, 74–77,
    83, 261
reefs, 194–95, 204
religion vs. science, 58–59, 60
Ricci, Michelangelo, 89–90
Richardson, Lewis Fry, 9, 134–40,
    251, 263
    Charney and, 187, 188–89,
        190–92
    death of, 192
    echo sounding and, 110–11
    on future forecasting, 217
    land features and, 155
    mathematical model of, 8–11,
        12, 20, 34, 41, 53, 109, 115,
        134, 135–43, 145, 157–59,
        176, 179–80, 184, 227, 254,
        256
    psychology and, 181–82
    resignation of, 180–81
    weather factory of, 9–11, 185–86
    World War I and, 115, 140, 141–42,
        178, 257, 266
    World War II and, 182, 205–6
*Rip Van Winkle* (ship), 43–44,
    50, 241
Robinson, John Thomas Romney,
    95–97, 138

*Robinson Crusoe* (Defoe), 65–66
*Rocinante* (sailboat), 3, 12–13
  animals around, 21, 25, 162–65,
    166–67
  in Belize City, 214–17
  casting off, 17–18
  collision course of, 70–72
  in Dry Tortugas, 160–61
  equipment on, 21–22, 87, 92–94,
    97–98, 110–11, 204–5, 207
  in Florida, 84–86
  forecasts on, 40–41, 53–55
  in Guatemala, 249–51, 253–54,
    256–57, 258
  horse latitudes and, 79–81
  in Isla Mujeres, 65–66, 170–72,
    173–75, 178–79, 181, 182, 185
  journey's end for, 262–66
  lee shore and, 198–99
  norther and, 25–26, 33, 36–38,
    42–43
  off Mayan coast, 193–95, 212–13
  in Placencia, 224–26, 229–30,
    233–35, 240–42, 244–48
  sails of, 31–32
  satellite over, 103–4
  in Venice, 115–16, 121, 125, 129,
    143–44, 146–47, 156–57
Rodgers, Calbraith Perry, 124, 125
Rommel, Erwin, 185
Roosevelt, Franklin D., 201
Rossby, Carl-Gustaf, 188, 189, 227, 266
Rossby waves, 187, 188–89, 227, 260
rotors, Flettner, 210–11, 214, 217, 266
*Royal Charter* (ship), 50–53, 55, 92, 245
Royal Meteorological Society, 96, 104
Rural Electrification Administration,
  201
Russ, G. L., 16

Saffir-Simpson scale, 172
*Sailing Alone Around the World*
  (Slocum), 209
sailing vessels, 207–12, 225–26
  ancient Egyptian, 196, 207
  clipper ship, 225
  ketch, 17

man-of-war, 27–28
  oil tanker, 213–14
  racing, 32–33
  SkySail, 211–12, 214
  See also *Rocinante;* ships
sails, 31–33, 208
  reefing, 204
  rotor, 211
  turbosails, 209–10
  wagons with, 199–200, 214
  windmills with, 196
Sajak, Pat, 130
saltation, 144, 148
sand dunes, 154, 155
Santa Ana winds, 23
satellites, 103–4, 106–9, 156, 252
Sawyer, Diane, 130
Saxby, Stephen Martin, 60, 61,
  116–17, 133
scatterometers, 108–9
science vs. religion, 58–59, 60
sea anchor, 29–30
*Shin Aitoku Maru* (ship), 213–14
ships
  apparent wind on, 93, 98, 110
  barometers on, 91, 92
  depth sounding on, 110–11
  lights of, 70–71, 72, 98
  sinking, 6–8, 43–45, 47–48,
    50–53, 55, 110, 241, 245
  steam-powered, 208–9, 225
  weather, 99–102
  See also sailing vessels
Skylab space station, 108
Slocum, Joshua, 209, 254
smartphones, 252
Smith, John, 25, 34
Smithsonian Institution, 76, 78, 86
sonar, 110–11, 170
Sophie, Duchess of Hohenberg, 140
sound waves, 110, 187, 189
spiders, 166–67, 168, 169
steam engine, 208–9, 225
*Storm, The* (Defoe), 4, 266
storms, 3, 19
  animal forecasts of, 165–66
  convection, 73

Defoe's, 3–8, 13, 266
dust, 149–51
    fear of, 16
    Franklin on, 38–40, 42
    in Placencia, 244–48, 249
    shipwrecks in, 6–8, 43–45, 47–48,
        50–53, 55
    theories of, 72–77, 78
    thunderstorms, 73–74, 264–66
    tornadoes in, 3, 24, 26, 206
    tracking, 43–47, 55, 57–58
    See also hurricanes
stratosphere, 105
suction pump, 87–88, 89
Swan, Lawrence W., 167–68,
        169, 170

Taino people, 24
Tate, William J., 122
Taylor, George, 7
telegraph, 45–47
    data network via, 91, 94, 115, 126
    forecasting and, 55–56, 57, 86–87
Teller, Edward, 230
theory, 251
    improving, 254
    observation and, 76, 78, 84, 109,
        127, 133
Thomas, William, 200
Thoreau, Henry David, 203, 204
THORPEX (The Observing System
        Research and Predictability
        Experiment), 258
thunderstorms, 73–74, 264–66
TIROS-1 (Television Infrared
        Observation Satellite-1), 107–8
Titanic (ship), 110
tornadoes, 3, 24, 26, 206
Torricelli, Evangelista, 19, 67, 87,
        89–90, 92
trade winds, 23, 162
    maps of, 38, 67–68, 98, 250
    reliability of, 26, 253–54
    theories about, 68–70, 77, 82–83
tree planting, 152
tropopause, 105
turbines, wind, 13, 214, 217

twilight, nautical vs. civil, 111–12
typhoons, 24, 29–31
    See also hurricanes

vacuum, Torricellian, 89
Vail, Albert, 45, 46, 86
Vanguard 2 (satellite), 106–7
ventifacts, 154, 155
Vidi, Lucien, 91
Voluntary Observing Ships program,
        101–2, 105, 156
von Neumann, John, 230, 232, 263
vortexes, 118–19, 153
Voss, John, 29–31, 171

Wanderer (ship), 44, 241
water
    air vs., 19–20, 66
    chlorine in, 176
    erosion by, 153–54
waterspout, 26
weather
    control of, 230–32
    statistics on, 50, 55, 63
Weather Advisor (ship), 100
Weather Book, The (FitzRoy), 63,
        165–66
Weather Bureau, US, 125, 146, 147,
        151, 206
Weather Channel, The, 131
weather data networks, 251–52
    aerial, 104–6, 124–25, 127, 140,
        152–53, 157
    improving, 254–56
    oceanic, 40–41, 67, 98–103, 105,
        112, 156, 252
    satellite, 103, 106–9, 156, 252
    telegraphic, 46–47, 55–56, 57,
        86–87, 91, 94, 115, 126
weather forecasting
    accuracy of, 64, 161, 217–18, 228,
        234–35, 252, 253–54, 255,
        257, 262
    by animals, 165–66
    astrological, 60–61, 117, 133
    aviation, 124–25, 152–53, 262–63
    barriers to early, 41–42

weather forecasting (*cont.*)
  Bjerknes's graphical, 84, 109,
    119–21, 125–29, 132, 135–36, 143,
    144–46, 160, 176, 185, 205, 227
  Charney's mathematical, 186–87,
    189–91, 222, 227, 230, 263
  commercialization of, 62
  computerized, 10–12, 134, 156,
    185–87, 189–92, 207, 217–23,
    234–35, 242–44, 255–56
  ensemble, 240, 242–44, 254
  for farmers, 128–29
  FitzRoy's intuitive, 55–60, 60–64,
    66, 84, 133, 207
  future of, 35, 217
  on Internet, 53–54, 131, 156,
    234–35, 260, 266
  Laplace's "if this, then that,"
    226–27, 232
  Lorenz's chaos theory and, 34,
    218–23, 226–29, 235–40, 244, 257
  military operations and, 180,
    183–85, 230–31
  models for, 53–55, 254–56
  on radio, 40
  research on, 259–62
  Richardson's mathematical, 8–11,
    12, 20, 34, 41, 53, 109, 115, 134,
    135–43, 145, 157–59, 176, 179–80,
    184, 227, 254, 256
  scientific theory in, 66–67
  storm prediction in, 43–47
  on television, 129–32
weather maps
  first, 38, 113
  fronts on, 115, 129–30, 146, 153
  indexing, 135
  isobars on, 37
  mathematics and, 121
  Smithsonian Institution's, 86
  synoptic, 38, 112–15
  wind speed and direction on, 54
*Weather Observer* (ship), 99
*Weather Prediction by Numerical
  Process* (Richardson), 8, 135–36,
    138–40, 157–58
weather systems, 113, 206

Weather Underground (website), 131,
  260, 266
Webb, Richard, 7
Welch, Raquel, 130
westerlies, 23
Wexler, Harry, 108
whistling, 79–80, 85
Wilberforce, Samuel, 58, 59
*Wild Wave* (ship), 44, 241
Willhofft, F. O., 211
wind, 19, 251–52
  apparent, 93, 98, 110
  Beaufort's scale for, 26, 27–29, 94,
    256
  direction of, 25–26, 54, 105, 138, 154
  energy from, 13, 196–97, 199, 200–
    204, 211
  erosion by, 148–51, 153–54
  Goldilocks, 22, 65, 162
  gusts of, 34, 246
  names for, 23–25, 48
  understanding, 20–21, 34, 67
  *See also* trade winds
wind barbs, 54
windmills, 196–97, 200–201
wind speed, 54
  global changes in, 260
  isobars estimating, 37
  measurement of, 28–29, 86, 93,
    94–98, 104–5, 108–9, 138
wind tunnel, 123, 124
wind wagons, 199–200, 214
women, on sailing vessels, 208
World Meteorological Organization,
  101–2, 105–6, 258
World War I
  Bjerknes and, 115, 127, 128, 132, 133
  gas warfare in, 177–78
  Richardson and, 115, 140, 141–42,
    178, 257, 266
World War II, 182, 187, 205–6
  D-day, 183–85
Wright, Orville, 122–24
Wright, Wilbur, 13, 122–24

yardangs, 27, 154–55
Yorke, James, 226